Lind
Christma
Lov
Grampa

ON THE BEAUTY OF SCIENCE

ON THE BEAUTY OF SCIENCE

A Nobel Laureate Reflects on the Universe, God, and the Nature of Discovery

HERBERT A. HAUPTMAN
As Told to and Edited by D. J. Grothe

Prometheus Books
59 John Glenn Drive
Amherst, New York 14228-2119

Published 2008 by Prometheus Books

Inquiries should be addressed to

Prometheus Books
59 John Glenn Drive
Amherst, New York 14228–2119
VOICE: 716–691–0133, ext. 210 FAX: 716–691–0137
WWW.PROMETHEUSBOOKS.COM

12 11 10 09 08 5 4 3 2 1

Library of Congress Cataloging-in-Publication Data pending

Printed in the United States of America on acid-free paper

CONTENTS

Does the scientist's world conform to the real one? Nature usually answers this question with an emphatic "No!" It has in fact been said that Nature delights in saying "No" and only with the greatest reluctance condescends to reveal her secrets. For this reason the scientist's life is not an easy one. However, on those rare occasions when his world does conform to the real one, and for this reason does throw light on the world around us, the rewards and the satisfactions are great and more than compensate for the many disappointments.

—Herbert A. Hauptman

Chapter 1

MY YOUTH AS A SCIENTIST

Though it has been a long time, I remember my childhood very clearly. In a way, I suppose I was really very lucky. My parents focused all their attention on their three children, their three sons. Their goal was for each of us get as complete an education as was possible. Of course, at the time I was growing up I had no way of knowing that this was different from any other family. But I do recall very clearly my parents' emphasis on education; to them the most important thing was to ensure that we three got as much education as possible.

I was born in the Bronx in 1917. Being born and brought up in the Bronx at that time was for me a lucky thing. In contrast to the situation of most parts of the country today, the Bronx had a marvelous library system. There were small libraries scattered throughout the borough, and these always were within short walking distances from where anyone lived. I went to the elementary school system in the Bronx, which I have since learned was an exceptionally good school system.

I was the kind of kid who just loved to read, and I read a lot. I developed an early interest in and read mostly science books, but not just science.

One of the public libraries I recall was founded by Andrew Carnegie, within walking distance of where my family lived. It took me maybe twenty minutes to walk there. Another library I remember was just five minutes away from where we lived. This library was just a storefront, a little store that had been made over into a library. If my recollection serves me right, this library was on Washington Avenue. I was there every weekend, every Saturday. This was before the advent of television; people could enjoy silent movies, and radio hadn't really begun yet. I think I would have still been interested in the library since I so liked to read. A book has sort of magic for me.

My mother had come from New York City, and her parents from Eastern Europe. My father was born in Austria, but he came to this country when he was four years old, I believe.

I was the oldest of the three boys.

I don't believe my brothers were great readers, so I suppose I was different in that way. I have never really thought to ask the question of why I was different, but there is no doubt about it, I was different. In fact, I can even remember a little trick that I tried to play, a scheme I tried to get away with: I wanted to get out a large number of books from the library, but I was allowed to take out only two at a time. I would try schemes so that I could check out more than I was allowed, but I was always unsuccessful; I usually tried to hide books in parts of the library I could later get access to. The plan was to get them later, but they were never where I hid them when I looked the next week.

Again, I was reading all kinds of books, not just books about science. I read what I suppose were the classics. I liked in particular the Russian novelists, such as Tolstoy. I read all of them. I read Thomas Mann's *The Magic Mountain*. I read many of Jack London's books. I began reading ravenously as early as I could follow what I was reading. I suppose the age that I began reading so much was around nine or ten, while I was still in grammar school. My grammar school was PS 57, on Crotona Avenue and 181st Street in the Bronx. We had moved when I was fairly young, then I went to PS 89 for the last year or so of grammar school.

I spent my last year of high school at Townsend-Harris High School. It was an exceptional high school, a three-year high school. Very few people could get accepted to this secondary school. I had to take an exam to get admitted. The reason I got into the school, I believe, is that my father had heard about it, and he wanted me to go to that high school.

It's funny, these things you remember: I remember having to take something like an IQ test at City College. I suppose this was because Townsend-Harris was sort of a preparatory high school for City College. The exam was given in the Great Hall. These days, City College likes to get me to come back there, and I have been a number of times, which I like doing—it is such a fine university. I remember being so impressed that first time when we students went in to take the IQ exam. I remember the exact words of the professor who was giving the exam. He started in by saying, "This is an exceedingly large hall." Those were his exact words. And he meant by that of course that we should be very quiet. He didn't want any noise, and so we took the exam in complete silence. As I recall, I got into Townsend-Harris in 1930. Therefore, that exam must have been given in the early part of 1930. From our school there were eleven who took the exam. Three of us passed it.

It amazes me the things one remembers. We came that day to school, entered our classroom. Miss Houseman was the teacher. When we came into our homeroom, written on the blackboard was the following statement—I remember it exactly—"If June grades are satisfactory, Townsend-Harris will accept Klein, Hauptman, and Simons." It's funny, this was the beginning of 1930 and I can still remember it very clearly.

This period of time was the beginning of the Great Depression, after the 1929 stock market crash. My family was largely protected from the horrible effects of the Depression because my father always had a job, even though he never earned a lot of money. He was a clerk at the post office when I was very young. I remember that his salary was twenty-eight dollars a week. At the time, a family could survive on that. I remember that rent was always around thirty-five dollars a month.

My mother also worked. She was a part-time ladies' hat saleswoman. Her friend and her friend's husband owned a millinery store. My mother worked most days. I remember I didn't like the idea of mother working, as I would come home and she was not home. I liked the idea of having my mother around. But coming home and being alone, I did develop very early a habit for studying, which I suppose was a good habit. I just got into the routine of it. I would come home, while in high school, and simply have a routine. I would do my homework sitting at the kitchen table, starting from when I got home at four-thirty or thereabouts, and work until suppertime, which was around six.

I was never distracted by my brothers; they had their own lives. They were not studious like I was. Both were younger. One, a year-and-a-half younger than I was, just loved baseball and knew all the teams. And my youngest brother, who was eight years younger, was just too young to be much of a distraction from my studies. My little brother ended up as a secretary for Helen Harpers, a company that sold ladies' dresses and clothing, and he worked with them until he retired. My youngest brother became an electronics engineer and had a reasonably decent career.

I remember that though my father was very loving, he was also very strict. I think sometimes my parents were just so busy surviving that they left us kids more or less alone, or at least me. I think my parents were reasonably typical. They were Jewish but were not at all religious, so I have come to whatever outlook I have on humankind's place in the universe honestly.

But I was influenced. I did so much reading, and my favorite author was Bertrand Russell. I read a lot of science and math, at least the math that I could understand at that age. But I read everything Bertrand Russell wrote. He just made good sense to me. It was good that I could read him and know that I wasn't all alone in this outlook based on the worldview of science.

In fact, I was never even bar mitzvahed, which was pretty unusual at the time. I think I wasn't bar mitzvahed by default, not by design. My father was the youngest of four children. We knew and visited his siblings, though they were not extremely close. However, my father and my cousin knew each other quite well, though there was a fair difference in ages, probably twelve or fourteen years. My cousin and my father pretty much had the same secular worldview, and they liked each other a great deal. Neither of them were at all religious, and I think for that reason the rest of the family sort of regarded them as strange, even if they weren't overly religious themselves. My father and my cousin saw each other rather frequently. My cousin's name was Hy, or Hyman. So I knew Hy very well, even though he was older than I was. It seemed to me that he was always going to go to medical school.

My family was never extremely interested in politics. I went to City College, and as you may know, City College was a hotbed of radicalism in those days. I was there at the height of it, during the Great Depression. But I was not greatly interested in politics. I was interested in going to school and doing what a student does. Even so, I couldn't completely avoid politics. Every day on campus there were speeches going on all over the place. They were picketing the college president's house—Robinson was his name, I believe. The students had all kinds of concerns and complaints, and they protested grievances such as that the college president didn't really care about them, that they should have had more of a voice in campus affairs. This student movement was very strong, even in the 1930s. What is quite remarkable is that the political movements and activities at that time left me pretty much untouched. I was interested in other things.

My interest in mathematics and science began as early as I can remember; in a sense I was hooked on doing science even as a child. I was reading science from the very beginning, certainly by the time I was around the age of eight or nine. Obviously at that age the science I read was very general and basic. Even so, I believe it was formative. Reading was very influential in piquing interests and passions that would sustain me my whole life. Possibly my interests at that early age plotted the trajectory of all of my future learning and scientific discovery. I had pretty much decided, even then, that I wanted to be a scientist. Even at that early age, I recall getting excited about the subject. I remember getting all those science books and reading them. I remember a four-volume book, *The Outline of Science*, by a man I believe was named Thompson. And in those days the books that you could find in the library were geared toward students, teenagers. So there were lots of pictures, which I just loved.

Math was one of my major interests from the very beginning, but astronomy is what I really got interested in at first. I remember at an early age making maps of the skies and also learning relatively early about trigonometry so that I could attempt to figure out the distance between different planets and so forth. I was stargazing by the age of ten or eleven.

I did it alone, not with my fellows. We lived in an apartment building, up on the fourth floor, and I could see some distance away, maybe six hundred feet away, that there was a great big smokestack. I thought one day that it would be nice if I could measure the distance from my vantage point to that smokestack. So I did it; it ended up being very simple. Our apartment was a two-bedroom apartment, and so we had a couple of windows facing out toward the smokestack. I could see through the bars on the fire escape and exploit them to solve the problem. I would triangulate—measure from one point to the next, twenty feet or whatever it was—then measure the distance, which was fairly straightforward. That was around eleven or twelve, I'd say. But what I really was proud of was measuring the diameter of the smokestack. I discovered that I could do that in a very neat way. By backing off a certain distance, I could enclose the smokestack itself within my vision between the two consecutive bars of the fire escape. And by measuring that distance and the distance from the bars to where I was, plus knowing the distance to the smokestack, I could measure the diameter.

When I went to City College I did have two very good friends, Henry Birnbaum and Herman Feshbach, who shared my interests in science and mathematics. We were hanging out all of the time, having similar interests and philosophical commitments, even though they were physicists and I was more focused on mathematics. Even so, I had an interest in physics. Both Henry and Herman are now deceased. Henry Birnbaum died in a tragic accident during the first year he was at Cornell; he tried to rescue some student who was having trouble in a lake and they both drowned. That was a major disaster. He was an only child and his mother called me up a few months later and asked me to visit her, which I did. And it was so sad. They were crying all the time that I was there. My other friend, Herman Feshbach, became a well-known physicist. He spent his career at MIT, where he had a distinguished professorship. They eventually named a center after him, but I unfortunately never got to see him after we went to college together.

Even now, I miss City College. I had such fun. Those students at City College were very bright; you couldn't get into City College unless you were outstanding. I imagine I got accepted because of my good work at Townsend-Harris. Very few students were admitted to City College; there was a very strong selection process. I went through City College getting As in math and physics, but Cs in everything else. Math and physics were extremely easy for me since it was something I enjoyed. People tend to thrive when they are able to do what they enjoy. I only worked hard and excelled in math and physics, and I had only one year in chemistry since it didn't really intrigue me.

Yet the Nobel Prize I was awarded was for my discoveries in chemistry. Even so, my discoveries were somewhat mathematical.

While at City College, I maintained my interests in the humanities, but I didn't really have much time to pursue those interests. The one non-math book that I do remember reading was a book by Morris Cohen and Ernest

Nagel, *Introduction to Logic and the Scientific Method*, which I imagine was a somewhat standard text. I didn't know Morris R. Cohen. Unfortunately, I didn't take his course. I took the course in logic during the summer time, but he was not teaching it then. As I mentioned, I had read virtually everything by Bertrand Russell. But I never formally studied much philosophy at City College.

I spent most of my time at City College with Henry and Herman. We were together all of the time, lunches together and so forth. We had pretty much the same academic schedule, although I got more into the math part of it as they got more into the physics. We would have these great, long conversations about science. We spoke about it a lot. I do remember over the summer vacations we didn't see very much of each other and then coming back in the fall we would always have discussions about the things we learned over the summer. We wanted to share with each other what we had been learning. We had some discussion about the content of courses, and we would discuss also the physics and math we were thinking about independently.

One professor stands out in my memory. He was by far the best teacher that I ever had. His name was Emil Post; he was a logician. I had him for this course in Theory of Functions of a Real Variable. He was an outstanding person and served as somewhat of a model for me. And he was brilliant, the best teacher I ever had, the best organized. He had planned out the lectures for the whole course, every lecture, three lectures a week for sixteen weeks. It was a very hard course; most readers will probably not have ever had any real-variable theory, I suppose. I remember when we took the final exam. I had done all of the problems up until the very end when I didn't have any more time. I think it was a three-hour exam. At the end of the exam, when I had to stop, I just wrote, "No More Time." I remember the comment he had made on my test when he graded it. I had a perfect grade for all of the problems except for the last one. He wrote, "That's too bad." He seemed to be very empathetic even though he didn't give me credit for the unfinished work.

Today many people are concerned that there's not enough public interest in science. Something in my childhood made science so important to me, but obviously this isn't the case with everybody. I have lately thought a great deal about why people seem less interested in science today. I really don't know why. I feel strongly that this is a topic that should be studied more. It may well be that learning is less of a quest, less exciting these days, which is paradoxical because of recent advances in knowledge. Maybe learning science is made so easy now, with TV and radio, that it is less exciting and less challenging. In my experience, it was the challenge of learning and doing science that was, in itself, rewarding.

I didn't get into science because of some public figure promoting science or because of programs designed to promote science to the public. There just seems to be something about some people that makes science interesting to them. In my case, not a single one of my teachers directly inspired me to pursue science. By the time I got to Emil Post, my love of and passion for science were already set. I was twenty years old by then. I simply liked to do science, and it's as simple as that. Darwin would understand.

Lucky for me, my parents left me to it, leaving me to my own devices, and they didn't overly try to push me in one direction or another. They left me alone to pursue my passions and created an environment that allowed me to flourish. My love for science is probably just a function of the way I am.

Maybe a sort of explanation could be given by Darwinism. Maybe it is a kind of specialization. Everyone has things he or she likes to do. Of course, what also made it easy for me is that I liked pictures and all of the recreational math projects I pursued. In my office here, I have a compound of five tetrahedra. As a child, I got a book of the five Platonic solids, and it not only had pictures, but it also had patterns to make these geometrical models out of paper. I remember seeing them and seeing and recognizing their beauty, their utter and impressive beauty. Once you see them, if you are the type of person I happen to be, you will simply *have to* love the Platonic solids. When I saw them, I felt compelled to construct them. I was a young boy, probably aged nine or ten, at this point.

This quality or character of my personality may not have been completely unique among the students of City College. Many students of City College became successful and competitive in many fields, making a great many achievements. Even so, I didn't see them as being exceptionally smart at the time because they were the only people I saw—I assumed that this is just the way people are, that all people are similarly driven and disposed.

Only when I graduated and saw what the real world was like, that most people lead uninteresting and passive lives, could I see how smart City College students were.

CITY COLLEGE NOBELISTS

At City College there are nine Nobel Prize winners, and City College is understandably very proud of that. And so they keep trying to get us to come back, and I have been returning fairly often. It impresses me that City College, which was not an elite school at that time, at least in comparison with the other major schools, had so many enormously talented students. Not only were there nine Nobel laureates from City College, including myself, but I believe that seven graduated from City College in the decade of the 1930s alone.

Jerry Karle, the chemist who shared the Nobel Prize with me, graduated from City College in the same year as I did, in 1937. He says that he remembers me from that period, but I don't recall meeting him. There was also a third Nobel Prize winner from that same year; so there were three of us in that year alone. There were an additional four other Nobel Prize winners who matriculated at City College in the 1930s. There were none before the 1930s, and after the 1930s, there was only one or two; one was just awarded to a City College alum relatively recently. I would have thought by now someone would make a study of why there were so many Nobel Prize winners from City College. My guess was that it was in the thirties, it was the Depression, and all of us were very serious students. We worked very hard, no doubt about it. I have thought that maybe the unusually high number has to do with the conditions of the Great Depression and that people were working hard to attain a certain social standing and achieve success. And possibly as a result, it just hasn't been the same at City College since then.

WORLD WAR II

I was in the war for three years, in the navy. I did not see battle personally, but I was exposed to it. I was overseas for a total of eighteen months and had been trained as a weather forecaster, but when I went overseas I did very little weather forecasting. For whatever reason, I almost never did the jobs I was trained to do.

I joined the navy, even though I had been exempt because I had been teaching electronics for the army air force. Teaching electronics made me exempt from the draft since it was deemed to be an important and hard-to-fill position. But in the middle of the war, in 1943, I enlisted in the navy because it seemed likely to me that I would be drafted eventually. I thought that, with my background, I could enlist as a naval officer, which, in fact, I did do. I thought I would become an electronics officer; instead, the navy sent me to weather forecasting school, which I ended up enjoying a great deal. So I went to New York University for nine months and became a weather forecaster. I then participated in additional training at Lakehurst, New Jersey. They then sent me overseas, where I actually did no weather forecasting, to the South Pacific. I toured all over the South Pacific, starting in the Admiralty Islands. At the end of my island hopping, I ended up in the Philippines. The war ended, and I was then sent back home. All in all, I was overseas for eighteen months.

Though I didn't really have any romantic interests while in college, by the time I was in the navy I was married to my lovely wife. I had gone to work for the Census Bureau in Washington in 1940, where I met and fell in love with my wife, Edith. It turns out, by interesting coincidence, that she and I had lived within half a block of one another, although we had never known each other in the Bronx. We met years later in Washington and were married in November of 1940. It was recently our sixty-fifth anniversary; she has helped make my life meaningful and as fulfilling as it has been, and I love her very much.

Edith was an elementary school teacher, a biology major, taking her degree from Hunter College. But when she graduated it was impossible for her to get a job, even a volunteer job. This was 1940. I graduated in 1937, went on for a master's degree, and then went on working with the 1940 census.

MY YOUTH AS A SCIENTIST

A NAVY NAIF

Now, getting back to my service in the war, I will admit that it wasn't very pleasant nor rewarding. I do have a little theory about why I survived. I was such an innocent young man. You hear stories of guys who get drunk then walk out into busy, oncoming traffic, and they don't get hit or hurt, but they don't get hit by accident. Well, I was incredibly naive as a young man, but no harm ever came to me, even unintentionally.

I was transferred from one unit to another incessantly, but in each and every case, we acted in supportive roles and were never on the front lines. They made me officer of the day once, which was well known to be a very unsafe job. One of the things that happened to me, for example, was that I was called up in the middle of the night due to an emergency. The shore patrolman called me up and said that there was a deserter in the camp. The deserter apparently had a girlfriend in the Philippines, and he deserted the camp every now and then to see her and to be with her. He had done this a few times in the past but had always come back to the camp. As long as you came back to the camp, they forgive you for "deserting" and there were never any really heavy repercussions. But this time he was away for three months, which made him an official deserter. Therefore, he was subject to heavy penalty. It became my job, as officer of the day, to apprehend him.

They told me that he was armed and dangerous. So, at two o'clock in the morning, the shore patrolman said an officer was needed to apprehend him and that I was the officer who had to do it. When someone is court-martialed in the armed services, an officer needs to officially do it. Now I had this .45, which I imagined, even at the time, was merely a symbol of my authority. I had never even learned how to use it. But I had it. I put it on. The shore patrolman and I went out into the dead of night; there were just a bunch of tents around, and we came to this guy's, this deserter's, tent. The shore patrolman suggested that I should be the man to go in the tent and capture him because I was the official authority, the officer. I then confessed to the shore patrolman that I didn't even know how to use the revolver, and he said he would make it easy for me: he put the cartridge in the gun for me so that all I would have to do was pull the trigger if it indeed became necessary for me to shoot the deserter during the arrest. The shore patrolman was obviously afraid of what might happen, so he stayed outside the tent even as he nudged me to go in the tent and seize the man. I flicked on the light quickly, and there was this poor fellow, fast asleep, and he didn't know what was happening. He was absolutely confused. I came up upon him with the revolver I didn't even know how to shoot. I slammed the .45 right on him, and I told him that he was under arrest. This poor guy had been fast asleep, scared to death as he had every right to be, being awakened almost violently in the middle of the night. He got quickly dressed and we took him out, marched him to the brig.

After this occurred, I asked the shore patrolman, "What am I supposed to do with this .45, now that it's loaded?" He said that there is only one thing to do; that once a gun is loaded, it cannot be unloaded. It had to be fired. He said we'd have to fire it. So he took it from me and pulled the trigger. Nothing happened. It just clicked. He had never actually loaded it in the first place.

I only once encountered the "enemy." With a kind of dumb luck one night, I didn't go to the movies when everyone else did. One lone Japanese pilot, just one man, flew over the camp and dropped a whole bunch of bombs while the movie was showing. A large number of soldiers were killed. And then he just flew away. He was never chased; nothing happened to the Japanese fighter pilot. There were many similar experiences I had, where my naivety and a kind of innocent luck insulated me from harm.

By far, the worst thing that ever happened to me in the military was when I was a fire marshal in the Philippines. One day there was an enormous explosion. This is just before the atom bombs were dropped over Hiroshima and Nagasaki. The explosion right outside of camp caused a mushroom cloud, a massive one, a kind which I had never seen before. The whole ground shook from the explosion. As fire marshal, I thought I should go down there and at least see what was happening.

The scene at the site of the explosion was total disaster. The only guy I knew there was the commanding officer, one Commander Mason. He rushed up to me, and he was obviously very upset by the disastrous situation. It was a major emergency and no one was there to assist. The explosion had occurred in broad daylight. Out on the pier, all hell had broken loose.

I said, "I suppose I should go out there and see what's going on."

And he said, "Yeah, I suppose you should."

He wasn't going to go out there with me; that was obvious. I jumped into my jeep and drove out to the explosion. I remember his last words as I drove away. He said, "Be careful, and don't expose yourself to the enemy." As I was driving out to the explosion, I could finally see what happened. There were dead people all around—arms, legs, and skulls—all over the place. I would say that over a hundred guys were atomized in that one instant of the explosion. Medics began to arrive to clean up the horror.

When I got out there, I found out what had happened: there had been a small boat, carrying about a hundred small bombs. There must have been a defective one, since the whole thing suddenly exploded. One of the bomb fragments had jumped across the water and landed on a barge, which was loaded with dozens and dozens of aviation tanks filled with hundred-octane gasoline. The shell of the barge happened to be very thin and when pierced exposed the fuel to ignition.

This gasoline became an incredibly massive fire, which somehow had to be put out. But I didn't know anything about putting out fires, let alone putting out that kind of a fire. Luckily for me, the camp next to us had a legitimate fire marshal, and he came driving up and explained that the fire was really going to be very tricky to put out. He said that it could be put out immediately with water, but by doing that, the shell of the barge, which was by that time extremely hot, would accumulate hazardous gases and then there would be another enormous explosion. His plan was to first cool the barge, which he did simply with water. He brought out a fire boat with a Chrysler pump and cooled the barge. Once he decided that it was sufficiently cooled, he then put the fire out and there was no explosion. Nothing further happened, except the horror of the hundred or so men who were killed.

As I mentioned, I was never commanding officer, but I was officer of the day. And I became fire marshal because when I came to the Philippines, I had been overseas about a year or so. Our commanding officer said that we needed a fire marshal at the camp, and he said to me, "We'll make you the fire marshal." I had absolutely no training.

He said to me, "That's nothing, don't worry. We're starting a school tomorrow and you will just have to go every day for a couple weeks."

I didn't do that, though. Here's why: I went the first day. I was in this class and was the only officer in the entire class. There were about a dozen other enlisted men, and we were all going to the fire school, all learning how to put out fires, all kinds of fires. The instructor was a petty officer. He said, "First we'll make a number of different fires. The first fire that we're going to make is as follows . . ."

There was an enormous cube, maybe eight or ten feet on an edge, which was hollow, but it was made out of iron or possibly steel. There was an entrance at the bottom end of the cube. The cube itself had been partially filled with diesel fuel. They put the whole thing on fire, ignited it. And then the instructor said that we had to put the fire out.

To put this fire out, we had to crawl into this cube, which was completely enclosed and on fire. There was this enormous cloud of black smoke, actually clouds of black smoke fuming from the fire. We had to put it out with water alone, with mist alone. The instructor said that since I was the only officer in the school, I would have to take in the enlisted men one at a time, going in there each time with every one of them individually. Now, try to imagine this: a fire inside this big cube, black smoke coming out, unbelievable amounts of thick, black smoke. And I would have to go in and guide each of the enlisted men in with me. Put out the fire. Come back out. Then they would start it on fire again and I would have to take the next student into the infernal cube.

So, in fact, I did this and it worked pretty well, at least for a while. I put out ten fires that way. And when I finally got out after doing this ten or twelve times, which took maybe two hours, I became very sick. I began to insist that I simply couldn't do it anymore. All that smoke gave me serious smoke poisoning. I simply could not do it anymore. The instructor finally said that it was okay, that I could rest for the remainder of the day, and that the class would finish the day's work with the other enlisted men. In fact, I never went back to the class. I felt so sick that I literally thought I was going to die. I was taken back to the camp and there checked into the hospital, staying for about ten days straight. By the time they let me out of the hospital, the fire school was over. Nonetheless, they had me remain as fire marshal.

So I had a number of jobs in the military, none of which I did particularly well.

MY YOUTH AS A SCIENTIST

MAINTAINING MY INTEREST IN SCIENCE

When I enlisted, I imagined I might have time on my hands while on the tour of duty to continue my scientific pursuits, or at the very least, to continue my reading while traveling overseas. So I had brought some books with me. One that I remember bringing was Wilson's book on advanced calculus. Whatever spare time I had, I went through that book. Not only did I go through that book, but I wrote notes on everything that I thought about regarding mathematics and science during the eighteen months I was overseas.

Chapter 2

THE CRYSTALLOGRAPHER'S CHALLENGE

UNIVERSITY OF MARYLAND AND THE NAVAL RESEARCH LAB

After the war, when I got out of the navy, I went to the University of Maryland for my PhD. The title of my thesis was "An N-Dimensional Euclidian Algorithm," which concerns certain aspects of number theory. My experiences as a student at the University of Maryland were very different from my experiences at City College. There was simply no comparing the students at Maryland with the students I knew at City College. Now of course, maybe this was because the students at the University of Maryland were all much younger than I was. I was the only student in the program who was a veteran. This must have been 1947. I was about thirty years old and, as a result, probably much more serious as a student than the others.

I also went back to my old job for a year or so, until I got a position at the Naval Research Lab, which was the kind of job I was looking for. I wanted to get a job doing science, doing some kind of scientific research, and I was very eager. I didn't care much at the time if the science I got involved in was theoretical or experimental. As it turned out, when I went to the Naval Research Lab and started working with Jerry Karle, who was my immediate supervisor, the first assignment that he gave me was to design and construct a slow-electron diffraction instrument. As with many of my experiences while in the navy, I knew next to nothing about what I was supposed to be doing, which in this case was electron diffraction. But I quickly learned. Jerry had been working at the lab for about a year or so when I came there and was already working in electron diffraction.

He had the idea, which was no doubt a good idea, that it would be worthwhile to study and work out how slow electrons are diffracted by gasses. If a beam of slow electrons strikes a gas and we look at the way the electrons are scattered as a result, we can discover some important information about the structure of that particular gas. I didn't know anything about it but was willing to learn. And I did learn. And not only did I learn but also I designed an instrument that would allow us to conduct the experiment. It took me maybe a year or so, then it had to be constructed.

I had not yet graduated from the University of Maryland. I was working while I was going to school. I liked the idea of going to the Naval Research Lab because they had a program that encouraged people to go to school while they were working, which was perfect for me. It enabled me to do scientific research, which seemed to satisfy a need I felt, and to go to school at the same time. At the time, I didn't even care too greatly what research area I was assigned to work in. I just wanted to get into scientific research. Of course, if I had continued forward there and gone ahead with that line of scientific inquiry, I would have become an experimentalist, not that I would have had any problem with that. During the year I spent designing this instrument and the additional year or so that it took for it to actually get constructed, I learned about x-ray diffraction.

X-ray diffraction is not exactly electron diffraction, but it is related. I learned also in those few months that there was a major problem in the field of crystallography, and this was lucky for me, proving to be a very satisfying challenge for me psychologically. The major problem that x-ray crystallographers were worried about at that time was what came to be known as the phase problem of x-ray crystallography.

X-ray diffraction is now the primary means for determining the structure of molecules, and it works by passing an x-ray through a molecule that has been crystallized. The pattern of the diffracted x-rays, which are recorded as an array of marks on a photographic plate, depends on the arrangement of the atoms in the molecule whose structure is being determined. The phase problem was that the diffraction pattern registers only the intensity of the waves. In order to solve the structure of the crystallized molecule, it's necessary to not just know the intensity of the waves, but to also know the relative timing when each wave hits the photographic plate, which is called phase data.

This was the problem that all the x-ray crystallographers were absolutely certain was unsolvable, even in principle. They all concluded that there was reason to believe that you simply could not solve this problem by working back to the structure of the crystal from the available data. Now I came on the scene, knowing nothing really too extensively at that point. All that I heard was "Here's a problem which is unsolvable." Well, I was a mere thirty years old or something like that, and I took up the challenge.

As such, I think that I had the needed psychological incentive for learning everything about x-ray diffraction because of the challenge posed by this phase problem of x-ray crystallography. This is why I was so good, along with the fact that I had such a strong math background; in practically no time, I had learned all the x-ray crystallography that I had to know in order to tackle this particular problem. I've worked on it from 1947 to the present time, but for those first eight years or so, it was day-and-night work, virtually nonstop. During this time, my wife and I had two children, but I was incredibly focused on my research. I doubt that I would have the stamina now for such strenuous research, but at that time I could do it, and with a deep satisfaction. I would even say a certain kind of joy. I stayed up until one or two o'clock in the morning almost every day, seven days a week; all the time, that is what I was doing. Luckily, my wife was a very helpful support with the children.

It was the challenge of the believed-to-be-insoluble phase problem that excited me. I mean, here was a problem that everyone was convinced had absolutely no solution. It didn't take me long to figure out that there was enough information, enough data from the diffraction data, to solve the problem and work back to the structure of the crystallized molecule. I became convinced that there was indeed a solution that could be discovered, that, indeed, the problem wasn't insoluble. The conviction the problem was not solvable is what bogged everybody else down; everyone felt that the information, the data, was simply not there to solve the problem.

But once I had formulated the problem as a purely mathematical problem, I could see that there had to be a solution. I did this easily, given my enthusiasm and background in mathematics. There was a mathematical relationship between the diffraction pattern of a crystal and its structure—when you shoot the beam of x-rays at a crystal, the crystal scatters the x-rays at different intensities in different directions. This gives you what is called the *diffraction pattern*, the directions and the intensities of the scattered x-rays. There are lots of them, literally thousands of scattered x-rays, whereas the number of scattered x-rays that were needed to determine the structure—to define the structure—was some finite number.

This means that there are only fifty or a hundred or two hundred unknowns. But the number of known observations could number in the thousands. What this meant, without even knowing any of the details exactly, was that the information necessary to solve the problem was, in fact, there in the diffraction pattern. It involved not simply a formula but an approach, a way of proceeding in order to solve the problem.

By the time our monograph, "Solution of the Phase Problem," was published, it was September of 1953. I wrote the monograph in July. Jerry Karle did not contribute to this paper. In my paper, we used x-rays to determine the structure of crystals instead of slow-electron diffraction through gasses, which had been Jerry's idea.

I worked solid the month of July, day and night, and did nothing else. I think that's the best way to work.

We were not awarded the prize until 1985. The work seemed to be largely ignored before that, which may be the case with many scientific breakthroughs. The reason that it did take so long for its importance to be recognized is that this work enables you to determine molecular structures routinely and rapidly, which only became significant and very useful later in the twentieth century.

You can determine molecular structures in a day or two today because of technological advances. But the first structure that we determined, the colemanite structure, took us several months. What came to be understood in the thirty years or so since the original monograph appeared is not only that one could determine molecular structures rapidly, but also that this work related structures of biological importance to their biological activity.

This work is now the foundation and basis for designing new drugs. The pharmaceutical corporations all have crystallography departments. They all have staff who specialize in nothing else but x-ray crystallography, but for bigger and ever-more complex molecules now. Modern medicine is galloping ahead at an ever-increasing pace, and in pharmacology, this is due in large part to x-ray crystallography and our "direct method." Because of our ability to determine molecular structures rapidly and accurately and to relate these molecular structures with biological activity, we get a better understanding of biological processes. We have a much better understanding of why some drugs do what they do and have adverse side effects and so on. So simply, we can design drugs in a more rational way than was ever possible before. As a result of our solving the phase problem, the development of drugs is now entering a phase of rational drug design, rather than drug discovery, which is the way it was before our direct method. It was always hit or miss before, and now there is nothing hit or miss about it.

Since the 1980s especially, as the method kept getting more and more powerful, it had advanced to the point where even protein structures could be determined. It was then that the Nobel Foundation said, "Look, this is very important stuff."

It has become important in other fields as well, not just in the life sciences but in the material sciences as well. You can relate structure with physical properties of materials. It is also important in mineralogy. When we solved the phase problem, of course, we had no idea it would become this important. As far as I was concerned, this was just a very challenging yet very satisfying problem to work on.

It is a paradox that our society seems to cherish our Nobel Prize winners and sees scientific research as crucial for the advancement of the human condition. But I was not inspired to work on the phase problem by the prospect of winning a Nobel Prize. There was great joy in simply doing the research and solving the problem. In a sense, it is really the only thing that I have ever really wanted to do. Research on the frontiers of science is the only thing I have ever really wanted to do.

REJECTION BY THE CRYSTALLOGRAPHY COMMUNITY

As I stated, all crystallographers thought solving the phase problem was out of the question, we were wasting our time, and the phase problem was simply unsolvable. A historian of science might be interested to hear some of the reactions we received when we gave our papers out at these crystallography meetings. They were saying, "These young kids, they don't know what they are talking about." We were absolutely dismissed. It was very hard emotionally, yet we persevered.

It took the crystallography community some time to see the merit of our solution. Recognition came well after I personally knew that we were on the right track, which I knew almost from the very beginning. But I didn't foresee just how hard it would be to solve the problem. It took at least eight years, from 1947 to 1955, for others to recognize the solution. Again, I knew well before then that this problem was solvable, which others denied. Again, how did I know? Because of this very simple argument that there were far more observables than there were unknowns. It was similar to solving a system of twenty equations and ten unknowns. You have more equations than you need to solve the problem. So I knew when the crystallographers were saying this problem was unsolvable, they were all wrong. They said it was unsolvable in principle. But I knew the information was there. And at least knowing that provided the impetus to keep searching for the solution. It is difficult for me to convey what it was like, the excitement connected with the sense of challenge.

INITIAL APPLICATIONS OF THE DISCOVERY

In 1953 a friend of ours at the US Geological Survey, Charlie Christ, had a problem. The problem was the structure of colemanite. Colemanite is a mineral, the structure of which was unknown at the time. And he didn't have any idea how to solve it, especially since it had a completely unknown kind of structure. So he came to us. We were, after all, claiming we could solve the structure of crystals with this new method.

Charlie Christ and his colleagues collected all the data at a couple thousand intensities to determine only about twenty atoms. It was a greatly overdetermined problem. They collected the data and we went about our method of finding the phases that were needed to calculate the structure. It took us about a month to determine the phases and calculate the structure by hand; Charlie Christ would come over to our place a couple times a week and we would do the arithmetic. We determined the values of a couple hundred phases, but we needed to determine the values of a couple thousand phases. So for this we had to write a very short computer program and to determine the values of a couple thousand phases.

We didn't have a computer up to the task, so we managed to use the computer at the Bureau of Standards. Peter O'Hara did the programming. We had to calculate what is called a Fourier series, which in those days took about a month to do. (Nowadays it takes about a minute, to give you an idea of how computers have progressed.) Then we had to plot this thing on a three-dimensional board, so we put together a big three-dimensional plotting board and started plotting these twenty or so points.

I will never forget the day that we did this. All of us were over at the Geological Survey. We began to plot these twenty or so points on a three-dimensional model. Howard Evans, a first-class crystallographer there, was looking at the model of the structure, the ways the atoms were shaping up on our three-dimensional board. He was standing and watching. At some point during our construction of the three-dimensional model he announced, "This is clearly the answer."

This structure consisted of five calcium atoms, maybe half a dozen silicon atoms, and then a bunch of oxygen atoms, possibly some carbon atoms, too. And we could see these two perfectly regular tetrahedra. And Howard Evans looked at that and said, "This is clearly the answer." And it was.

It feels impossible to describe how seeing the solution take form in the three-dimensional model made me feel. I knew then that this was working and that this was a breakthrough. We knew that it worked. I knew it, Jerry Karle knew it. The others could see it working. They could see this structure appearing out of nowhere as we plotted the points on the three-dimensional board. The colemanite structure has now become kind of classic. This guy Coleman, after whom the mineral is named, actually made a sculpture out of the solution. The sculpture was given to me as a gift by Charlie's widow after getting the Nobel Prize.

From that moment on, people were still not believing this was the solution we were claiming it to be. And so it took another five years or so, until 1960, when two things happened. First, Jerry Karle's wife decided when she saw our solution that she would get into the business of using the method to determine crystal structures, and she did that. Second, a crystallographer named Michael Woolfson, one of our main academic critics who claimed that what we were doing couldn't be done, made a complete reversal. He could see the method was indeed working, and he spent the rest of his life developing the methods, writing computer programs. He was out of York University in Britain.

Michael actually became a very good friend of mine. I liked him a lot. And he did more than come on board with the idea. I don't know that he himself wrote computer programs, but he had people working for him who wrote the programs. He then gave the programs out to anyone who wanted them, for free. This was a version of open-source sharing of knowledge to advance scientific inquiry. Sadly, that spirit is waning in the business of science today. So starting around 1960, these programs were propagated all around the world, and people started adopting them very rapidly.

Woolfson was a naysayer at first, and another guy, Pepinsky. In fact, their paper was called "Have Hauptman and Karle Solved the Phase Problem?" and they answered absolutely no.

In a sense, all of the criticism and naysaying of Woolfson and Pepinsky and others was depressing. But we were absolutely certain we were right. It is interesting to note that what prevailed in the end was not that our argument in the journals was persuasive to the other scientists, but that people began using our method in the real world. That's how everyone came to know it worked.

NAVAL RESEARCH LAB AND INDEPENDENT SCIENCE

Luckily for us, working at the Naval Research Lab earlier on didn't require that we did research in a specific field. It did not matter what we were doing, as long as we were productive. In fact, I am not sure they were very interested in what we were doing, but they let us do it. It was an ideal place for us, even though in the end, in 1970 when I left there to come to what has become the Hauptman-Woodward Institute, the navy discovered that the work I was doing all those years had absolutely no relationship to naval problems. They finally wanted to redirect me in ways that I did not like.

The director of the institute at that time was Dorita Norton. She had been trying to get me to come here for many years, and finally I had a reason to come, since I was being directed away from my independent research at the Naval Research Lab; the navy was no longer happy to support my research. It turned out to be an exceedingly good thing that I came to the institute, except that Dorita, who was a dear friend, died within a year or two. All of a sudden, I became the research director, which I wasn't very happy about. She had initially promised, in order to get me to come here, that I can do whatever I wanted to do—study and research according to my interests—and never have to worry about anything else. When she died, I had to assume the role as research director, having to write my own research grants, which took time away from doing pure research.

Chapter 3

THE IMPORTANCE OF INDEPENDENT RESEARCH

The Hauptman-Woodward Institute got its start about fifty years ago as a result of the work and dream of Dr. George Koepf. He was an endocrinologist, and he had a patient he was treating. She wanted to show her gratitude to him because he had helped her a lot. Dr. Koepf was convinced from early on that basic biomedical research was the key to improvement of optimal health and increased quality of life. The initial goal of HWI was a public health one.

Koepf wanted to support basic research that he felt was the underpinning of improvements in the delivery of healthcare, and when Helen Woodward, a wealthy patient of his, volunteered to show her thanks to him for helping her, he suggested to her to that she endow a biomedical research institute. Hauptman-Woodward Institute is the more recent name; originally it was called Medical Foundation of Buffalo. To the best of my knowledge, the institute was founded with Helen's $3 million contribution, which was quite substantial for the day.

The history of HWI is somewhat complex because Dr. Koepf was not only interested in basic biomedical research; he was also a practicing physician. He was affiliated with the Buffalo Medical Group, and in fact he was one of its founders, as well. Both the Buffalo Medical Group and the Medical Foundation of Buffalo were housed in the same building, on the corner of Delaware Avenue and West Utica Street, largely built with support from the initial endowment that Helen Woodward Rivas donated. I believe that the National Institutes of Health also contributed. The research initially was basic medical research in endocrinology, the branch of medicine that deals with endocrine glands and hormones.

I met Dorita Norton simply by attending the meetings of the crystallographic society. Since there were probably only two or three research scientists at the institute earlier on, along with a similar number of technicians, Dorita had wanted me to join the institute several years before I actually made the transition from the Naval Research Lab. I resisted coming to Buffalo because I was very well established at the Naval Research Lab, where we made our initial discoveries solving the phase problem of x-ray crystallography.

I initially had a lot of latitude at the Naval Research Lab, which I came to in 1947, some twenty-three years before I came to the institute. As I have stated elsewhere, my initial job was to design and construct a slow-electron diffraction instrument, which I did in the first year I was there. This started everything for me, but I didn't actually work in the field of slow-electron diffraction for a very simple reason: it took me about a year to learn enough about electron diffraction to be able to design an instrument, and it was being constructed at the Navel Research Laboratory, which took about an additional year.

During that time, I had learned enough about slow-electron diffraction that I could have carried out some experiments in that field. But instead of doing that, I got interested in the problem that everyone was interested in at that time: the phase problem of x-ray crystallography. X-ray diffraction is not exactly the same thing as electron diffraction. I had learned just enough so that I could understand the relationship of electron diffraction to x-ray diffraction. You get much more information from x-ray diffraction than you do from electron diffraction. During the time—maybe ten months or a year while the electron diffraction instrument was being constructed—I had a chance to learn about the closely related field of x-ray crystallography and to learn also about the exciting phase problem.

It was generally believed to be unsolvable. There had been a very convincing argument to this effect, but by reformulating the problem—reenvisioning it as a purely mathematical problem—it was possible for me to see that the problem of phase determination, far from being an unsolvable one, even in principle, was a greatly overdetermined problem and solvable in principle. I think this is the first major contribution that I made. There was much more information available to the scientists working on the problem than the minimum amount needed to solve the problem; of course, you needed a lot of extra information because the information came from experiment, and experiment is always subject to error.

I moved from the group of scientists with whom I was working, such as Jerry Karle, to another part of the Naval Research Lab. I was moved to what was a mathematical section of the Naval Research Lab, run by Howard Trent, who was a mathematician and engineer. He was a very nice guy and we worked very well together. Unfortunately he died after seven or eight years or so, and his mathematical research group sort of broke up. I was transferred to another research group, this time focusing on solid-state physics. Changing the focus of my research was not at my own initiative, so I welcomed the opportunity to eventually move on to what became HWI, where I could continue my research without having to change its focus.

Until this time, I pretty much determined the work I would focus on. Bureaucrats and administrators at the Naval Research Lab never really decided, at least early on, who researched what. I was very lucky in this respect. I suppose I had such latitude because the work I had been doing the previous thirteen years was very productive; we did a lot of good work and it was published. I should admit that the work I was doing was likely not very useful to the Naval Research Lab. If one were strictly to formulate the mission of the Naval Research Lab, the work I was doing had very little application to the US Navy's needs. They were funding science for science's sake.

By the late sixties, my status in the scientific community had somewhat changed. I was well known in the crystallographic community by this time, and this was in large part the reason that Dorita Norton wanted me to come to Buffalo. She felt that I added strength to the medical research of the institute. It is interesting to note that in the late sixties, here was a female scientist in a leadership position. There were women crystallographers, but no others that I knew of in leadership positions. (I think that today there are more women in x-ray crystallography than there have ever been before.)

When I arrived at the Medical Foundation of Buffalo, it was growing. There were research scientists and technicians who assisted the scientists, like a nurse or a physician is to a doctor.

Dorita Norton died shortly after I arrived, which was personally devastating, losing a close friend. Her death also had a very serious impact on the medical foundation; Dorita was the principal investigator on a major grant we had with the National Institutes of Health, the so-called Steroid Grant, on which Bill Duax was doing a lot of work, as well. One of the first things we had to do was to see if we could transfer the role of principal investigator from Dorita to Bill, and the NIH was very understanding. When we told them what the situation was, they said it wouldn't be a problem. So Bill inherited the Steroid Grant.

This was around 1972. I became very good friends with George Koepf, the president of the medical foundation. For a little over a decade, I was doing pure scientific research until he decided he wanted to step down and he wanted me to be the president, which happened, although I resisted it. This was 1985, the same year I was awarded the Nobel Prize in Chemistry for solving the phase problem. During the years from when I initially arrived until 1985, George consistently showed a great deal of confidence in my work; he seemed to feel that I could do anything, and he would tell anyone that he met that I was a future Noble Prize winner. This made me feel very uncomfortable because I believed there was no way this was going to happen. Many years later, after he died, which was only a short time after I won the prize, I spoke to his widow about it. She of course was very proud of George. I asked how he could have known or if she knew why he kept on saying that I was going to win. She surprised me with the answer: she said that Dorita Norton had told George Koepf way back in the early 1970s that she was certain I was going to be awarded the prize some day.

A CHANGING INSTITUTION

The Medical Foundation of Buffalo, as I stated, began with research in endocrinology. But when I came here it had already shifted its research focus to crystallography. The primary reason for this was that Dorita Norton had been here. The change in focus is not due to it following the money that was out there in grants, but due to its following Dorita Norton's own personal research interests. She intentionally recruited other crystallographers such as Bill Duax, Vivian Cody, and myself.

THE IMPORTANCE OF INDEPENDENT RESEARCH

One might ask, just as the navy asked when I worked at the Naval Research Lab, what does crystallography have to do with the mission of the Medical Foundation of Buffalo? That has turned out to be of the greatest importance, because the technique of x-ray crystallography, once the phase problem was solved and it became routine to determine molecular structures using the technique of x-ray crystallography, turned out to have a major impact on biomedical science. With the ability to quickly and routinely determine molecular structures of biological importance (like hormones or drugs), researchers gained a much deeper understanding of life processes. We now understand, or are beginning to understand, how living things work in a way never before possible.

It is said that virtually any new drug introduced since 1975 has been determined using the technique of x-ray crystallography.

The practice of science, of scientific discovery, in both its scale and its scope, has changed an amazing amount in the twentieth century, as did the relationship of science to government and science to the universities. The universities have been the major institutions through which scientific research is accomplished, but HWI is not formally attached to a university; it does not get university funding nor grant degrees. It is more closely related to the University of Buffalo now than ever before. Historically, the science done at HWI was funded by grants from the government. The State University of New York at Buffalo's chemistry department was headed by Bill Knockings. He arrived in Buffalo about the same time I did and was a faculty member in the chemistry department at SUNY Buffalo. When I came here in 1970, I was immediately appointed an adjunct professor in the biophysical sciences.

I did some teaching; it seems to me every year or two I taught a course for the first five or six years I was here, and I have been loosely connected ever since. But in more recent times, ever since I received the Noble Prize, the university decided that once and for all I should be more closely connected. They have been paying one-third of my salary ever since that time. Up until that time, I was an adjunct professor with responsibilities, teaching and lecturing at the university. I taught the direct methods of x-ray crystallography for several years, and then they asked me to do other lectures. Now the Hauptman-Woodward Institute is the Structural Biology department of UB, yet it is still independent. This has been a very complicated issue, and that's why it took so long to establish the relationship. We started talking to the university. The university has picked up part of the salaries of maybe some five or six of us.

"INDEPENDENT" SCIENCE

While the Hauptman-Woodward Institute is independent, we are dependent on money. All scientific research is dependent on its funding. When we say we're independent, it is sort of a euphemism. We are dependent on money, but no one but ourselves is directing our research. In order to get funding, we have to make a credible claim that we can do such and such, we are planning to do such and such, and we are expecting to get such and such results. The National Institutes of Health and the National Science Foundation have their own grant-awarding programs; they want to see research done in such and such an area, and they request proposals from individuals to submit some sort of an application.

Along with six other institutions, we were recently awarded a $17 million grant for a protein-structure initiative from the National Institutes of Health, in fact. This research has to do with rapid throughput of protein-structure determination. Only one facet of protein-structure determination has to do with the methodology; there are other important aspects of it that have developed. With the advent of the methods, which we are largely responsible for discovering and devising, the problem has sort of shifted. It had been the methods; now it's in large part the crystallization. Proteins are difficult to crystallize.

The question faces us: In a democracy, who should be the ones deciding how tax money should be spent on scientific research? Who should decide exactly what kinds of research are engaged in? Some people who talk about "public science" say the government should not be in the business of deciding research, they should just be funding research. They are of the view that scientists should be deciding what to areas to research. And even the government, in an attempt to answer this exact kind of question, has looked to the best scientists for scientific advice; they hire the best scientists they can find.

Or at least they did until recently. It is well documented how our current government may actually thwart scientific research. In some very measurable ways, the current administration is the worst that science has ever seen. The Union of Concerned Scientists and the National Academy of Sciences (in the "Rising above the Gathering Storm" report) have recently made this point rather persuasively, as have scores of science journalists, notably, Chris Mooney in his *Republican War on Science*.

There are increased funding cuts; look what's happened to Roswell Park Cancer Research Institute of Buffalo, New York. I was told just a couple of weeks ago that a major grant that they have been awarded from the National Institutes of Health has been arbitrarily cut by a tremendous percentage, for no other reason than because the government had changed its priorities. The NIH budget, although it has not actually been reduced, has remained pretty much at the same level in recent years, and so, in effect, this amounts to a reduction due to inflation. And this is in strong contrast to the NIH budget for the last five or six years, which has doubled, and it's now flat. Roswell Park is suffering a major reduction now to the tune of millions of dollars.

When you ask me should we be spending whatever it is, something approaching $600 billion by 2010 or more, just to kill people in Iraq, or should we maintain the NIH budget, the answer is obvious. In the last analysis, in a democracy the people should decide, and invariably they do decide; they elect officials, but *the people* can make mistakes, obviously. They are not infallible, as is proven by recent elections.

This is why it is so important for the general public to have a better understanding of where science is headed, or could be headed with the right funding. This is why people should get educated about what science is capable of doing and the amazing benefits new lines of research may afford humanity. Only then can they elect knowledgeable representatives capable of carrying out their wishes. Numerous polls show that the general public very much likes the idea of supporting biomedical research; they want to see the funding increased. But then they are often dissuaded by these by elected officials who do not share the scientific values that would benefit our society.

I think in the last analysis, the people are responsible for making these decisions, and to do this they should be well informed, much better informed than they are now, especially given the profound social impact of certain kinds of research. The public needs to be better informed about science, about the promise new lines of research holds such as stem-cell research or therapeutic cloning, and why they need to get involved in the decision-making process.

The first way to do this, of course, is to learn science, even as a nonscientist. I would suggest to someone interested in the future of our country that they take a night class in "science appreciation" or do some reading in science. Incidentally, a great benefit to this, because learning science really means learning methods of critical thinking, will be to kind of inoculate people from being overly credulous and gullible. Our culture is filled with examples of people who promulgate untested claims, and the general public is taken in due to its lack of scientific understanding. Examples include some forms of alternative medicine, belief in psychics and ghosts, and, of course, religion.

A central problem with regard to this public gullibility and lack of scientific understanding among the public has to do with parents of very young children. Many children are brought up in an overtly antiscientific atmosphere. There are aspects of American culture that are not just nonscientific—not just *ignorant* of science—but there is actually a hostility toward science. The first thing to do is to prevent that from happening. Now, how can you do that? This is why I say I think this business of religion is the worst thing that could have ever happened to the human race, with people believing things for no reason at all. Would you believe, for example, that I am absolutely convinced because I had a vision of this that there are seventy pink elephants revolving around the earth at seventy thousand miles an hour? Would you believe something like that? Obviously, you wouldn't. So why do you believe these statements that are made by these priests and other religious leaders, similarly incredible statements, for which there is no basis and that certainly cannot possibly be true.

I believe that from an early age most children in our society are inculcated in superstition and mumbo-jumbo, and so there is no development of the scientific approach to looking at the world. I believe there is a direct negative connection between belief in religion, especially fundamentalist religion, and public scientific illiteracy. And public scientific illiteracy directly affects the funding of good science.

Consider the Bible. It has influenced, for example, a favorite piece of music of mine, Handel's *Messiah*. I am deeply moved by it, and I listen to it knowing full well that I can't believe a word of it. But who cares? It is beautiful. This is similar to the Bible. Taught as literature, as a work of art, the Bible is fine and can be appreciated, or at least parts of it. I think religion is only fine if you treat it like other works of art and you suspend your disbelief while appreciating it. The *Messiah* gives me enormous pleasure, although I know full well that it is not to be literally believed, so obviously it must be possible to find pleasure in certain aspects of religion without believing a word of it.

Chapter 4

HOW GOD HURTS SCIENCE

Over the last year, I have been more vocal about some of my views regarding the compatibility of religion and science, causing some controversy. Despite the controversy, I feel compelled to say that I'm convinced science and religion are absolutely incompatible. While I never tried to pretend my views regarding religion are anything other than what they are, I've never really advertised them in the past. The issue of religion came up almost never in my entire life. It's only been in the past year or so that I have begun to speak out in ways I never really have before, all as a result of this article that appeared in the *New York Times*.

Science and religion are simply two different ways of looking at the world. When seeing the universe from a scientific outlook, one only believes that for which there is some evidence. With religion, on the other hand, one believes on no basis whatsoever; religion is just a matter of faith. The religious person believes the most ridiculous things for no reason. Most religious claims have no justification, and it seems to me it is not even open for discussion for most people. Most religious people cannot even stand to have their beliefs examined, much less criticized. Bertrand Russell has a good point to make in this regard. He says something to the effect that if you consistently apply the scientific approach to looking at the world and only believe things for which there is adequate evidence, this would necessarily overturn some of the most fundamental beliefs in our society, especially belief in the impossible claims of religion.

There was a meeting of City College Nobel laureates, a half dozen or so. City College is very proud if its Nobel laureates and has I believe about nine of them by now. And about five or six of us were invited to this scientific meeting last year, in which the contributors to the scientific sessions were students, and not only students of City College but New York City college students in general. The students gave presentations that we Nobel laureates judged and assessed. But we were there primarily because the organizers of this meeting had planned a panel discussion in which we would address certain scientific topics in front of the student body and the public.

To my surprise—and I think to everyone's surprise—the first issue that came up was raised by one of the students who asked the question "Are science and religion compatible?" This topic is something about which I've had strong beliefs now for about seventy years, and so I thought I would give my view. I responded to the question by saying something to the effect of "absolutely not, they're not compatible" and gave some initial argument for the position: science is based on evidence, and religion is based on faith, which is, by definition, the lack of evidence. So it seemed to me there is just no question about it, and I stated it very emphatically.

The initial reaction among the audience was essentially very muted; in fact, none of my Nobel Prize–winning colleagues responded at all, neither in support of what I said nor in opposition to it. There was initially no further discussion about the issue. Afterward, two or three of them did tell me privately that they supported my view. But for whatever personal reasons they had, they did not speak out at the event.

It is controversial to speak out against religion in our society, and the scientists who were there were not looking for controversy. They're an older bunch and I supposed they didn't consider it important enough. Also, I honestly think there is a matter of fear, fear of actual physical harm befalling an outspoken critic of religion in our society. Nevertheless, I spoke out without really thinking about it. There seems to be almost nothing more controversial to talk about today than religion. This is one reason it should be talked about more—we, as a society, should be open to examining our beliefs and open to changing them if we find there is no good reason for holding them. Still, for whatever reasons, my colleagues neither publicly supported nor opposed my position regarding the incompatibility of religion and science.

Attending the meeting was a reporter from the *New York Times* named Cornelia Dean, who was then an editor of the science section. She was something of a moderator for the event at City College. During the question-and-answer segment of the program, she was the one who expressed opposition to my view. We disagreed and I thought that was that, not thinking much of the whole affair. We dispersed and that, so far as I was concerned, was the end of it. But it turned out not to be the end: two or three weeks later I got a call from the *New York Times* science reporter, this same Cornelia Dean. She wanted to interview me on the phone about my comments made at the City College panel discussion, and our conversation lasted about a half hour. She responded on the phone to my views in an interesting way.

She asked me if I knew any scientists who were religious. She asked me if I believed that most scientists were not religious. Now of course there are religious scientists and scientists who are not religious, but in what exact numbers I did not really know, and so I felt it was a rather inconclusive conversation. At the end of the conversation she repeated that she was the science reporter from the *New York Times* and that she would use this conversation in an article she was writing.

This all started me thinking more about the subject. I decided to try and find out if scientists tend to be religious by initially asking people I work with at the Hauptman-Woodward Institute. To my great surprise, most of my colleagues and friends at HWI are religious. We had never really spoken about religion before, with possibly one or two exceptions. Most scientists are content to just do science and do not spend a majority of their time talking about religion.

I recall one colleague whose views regarding religion I asked about; he said he was religious, and I told him why I thought religion made no sense to me, and I asked him why he believed what he believed, and he gave me an answer. The answer was something to the effect of all he has to do is go out at night and look up at the sky and see all those stars and he knows there is a God up there. It was a very friendly conversation.

In any event, the article did appear in the *New York Times*, and it caused a stir, generating many e-mails and letters to me. They were all supportive and in agreement with me, with one strong exception. I found it interesting that letters were written to me both from general readers of the *New York Times* and from people in the scientific community.

THOUGHTS ON THE RELIGIOUS SCIENTIST

I think there is no doubt that science and religion are incompatible. This is not the same thing as saying, however, that scientists cannot be religious. Of course you can be religious and be a scientist—a person can indeed be inconsistent. Consider Giordano Bruno, who was consistent to the end, and it didn't help him survive. And on the other hand, think about Galileo, who lived around the same time, and recanted—was inconsistent—and he did survive. Inconsistency sometimes then has a survival value.

Newton was religious. And so was Einstein, in a sense. In fact, Einstein was religious in the way that I myself might be called religious, which is to say that there appears to be some sort of logical structure that governs the nature of the universe. But order in the universe does not imply a designer, and using the term "religious" to describe a recognition of apparent order in the universe is not very helpful and indeed may be confusing. The term religion implies belief in the supernatural, which I do not hold, and by all accounts neither did Einstein.

Scientific knowledge is admittedly incomplete, and this should not scare us. We certainly do not have to complete our knowledge by appealing to supernatural causes. I imagine that science will continue explaining the universe in natural terms, growing the body of knowledge we have about the universe, without recourse to supernatural explanations, such as God. That the universe is capable of being described by means of a mathematical structure is remarkable, but it is not supernatural, and doesn't seem to me at all to be evidence of the existence of God.

The universe is here, and that fact is itself utterly amazing. I'm certainly not the first one to have observed that. Einstein's wonder at the universe, which he used the term "religious" to describe at times, is just this appreciation of the laws of nature. The fact that it can be described by the means of a logical, mathematical structure doesn't mean that God is a mathematician—it does not follow.

So even if there are no logical reasons to be religious, there are so many social reasons to be religious. Religion can provide a social network of support, especially if there are no secular alternatives. Also take, for instance, if your wife is religious, and your family is religious, and your community is religious, and every Sunday you go to church—you may seem religious, even if you are personally a nonbeliever.

That people are religious at all used to puzzle me much more than it puzzles me now. I see this as Darwinism at work. It seems like people have these kinds of unfounded beliefs almost genetically, just like, for example, I love working on mathematical problems. For me it's very interesting to solve these mathematical problems, to work on these solutions, and it is the way I am made up. Indeed, there may be a genetic component. There is an enormous amount of pleasure for me in discovering something new. Most recently I have done this work on conic sections, and I've discovered a few things. I think I am hardwired to enjoy this type of activity. Other people seem to be predisposed to being religious.

WAS I BORN AN ATHEIST?

No one is born a believer in God—people are taught to believe in religion by their parents—so I suppose by definition, I was born an atheist. But I think my beliefs about God actually resulted from a great deal of reading that I did as a teenager. It was until I was about twelve or thirteen years of age that I was actually a very religious person. Now, I wasn't Jewish, per se, but as a child I believed in the supernatural. I believed that there must be a superpower out there because everything in existence is so marvelous. I mean, the world is amazing to contemplate, and doing so can stir the chest, making one feel religious. I later concluded that to assume a supernatural power exists is no explanation at all, because it simply transfers the problem of origins from here to there.

I also came later to see that the theory of evolution by natural selection, Darwin's theory, explains the origins of life—from very simple beginnings to amazing complexity—without any need for the supernatural, which itself cannot be tested, looked into, nor examined like scientific claims. Darwin's theory, on the other hand, is testable and passes the test.

Even in college I never really had long conversations about religion. By that time I was focused on science and spent most of my time talking with my two best friends about scientific matters. I never really knew if they were religious, but I would suppose that they were not.

I do find it encouraging that the top scientists in the world, members of the National Academy of Science, for instance, are atheistic. Something like 60 percent of all scientists are nonreligious, secular, or atheist, but over 90 percent of the top scientists lack belief in God. Some Nobel laureates have spoken against blind faith and unfounded religious belief: Steven Weinberg, James Watson, Francis Crick.

On the other hand, there are religious scientists. Take Francis Collins, for instance. He is involved in the Human Genome Project. In fact, he is mentioned in the *New York Times* article that discusses my views on this topic. Collins sees no necessary conflict between science and religion, and Stephen J. Gould held a similar view, although he himself was an atheist.

WHY DO PEOPLE BELIEVE IN THE UNBELIEVABLE?

It is very hard for people to accept that there may be no ultimate meaning to life. There is meaning, I admit, but I do not think an ultimate meaning. Even if there is no God and life is ultimately meaningless, you don't end up loving your wife any less than you would if there were a God. Most people seem to think there must be ultimate meaning to life. I really don't see that there needs to be. In fact, I'm not sure in what sense that statement makes any sense at all. I don't know what kind of an answer you can expect from inquiring into the ultimate meaning of the universe.

Sometimes you hear religious people say that if you don't believe in God, then you must be a rotten person. They ask atheists what keeps them from stealing from other people and murdering and raping and pillaging. They conclude that there is no reason to be a decent human being if there is no God. But if that is the reason you believe in God, because you are afraid of what might happen if you aren't a decent person, that's not a very good reason either. Why be good? Simply because this is the way some of us are. Some of us are good, others are not. And evolution explains this.

Sixty or seventy years ago I read Jack London's books, such as *Call of the Wild* and *White Fang*. In one of those books, I think it was in *Michael, Brother of Jerry*, there is a sentence that illustrates this point: "The horse abases the base, ennobles the noble. Likewise the dog." What that means is that a person, a dog, or a horse, etc., brings out the best in a good person and the worst in a bad person. This one sentence impressed me a lot, although I haven't read it in, I don't know, sixty years or seventy years or more. I still remember this one sentence today.

I would suggest that major factors determining people's ethical behavior have to do with their childhood and child rearing. I believe that I was rewarded for certain behaviors as a child that, in turn, inculcated certain habits. At a basic level, I participated in a reward system that rewarded me for working hard and being a decent person. Sadly, not everyone has this kind of environment when growing up. Religion may actually stifle this kind of upbringing.

SECULAR HUMANISM

If the term "secular" means nonreligious, and the term "humanist" means a decent person who puts humanity's well-being as a priority, then I would say I am a secular humanist. I suppose that I am also the kind of person who would be contemptuously referred to by those on the Far Right as a bleeding-heart liberal. I find myself just feeling sorry for many people and wishing to help them. There is lot of suffering in this world, and this suffering affects some of us more than others. Some people suffer more than others, and some people are moved by the suffering of their neighbor more than others. The suffering of others happens to affect me deeply. I find it interesting that those who say they are motivated by an ethic of Jesus Christ often behave in very uncaring ways toward the less fortunate in our society, and to me, seem less moved by human suffering. Yet, without any expectation of reward in heaven or fear of punishment in hell, secular humanists are moved by the inhumanity of man to reduce human suffering in the world. Indeed, I think that the belief in heaven and hell may inhibit the moral development in some people because it removes the impetus for some people to make life better in this world, in the here and now.

SCIENCE AS A METHOD OR AS AN OUTLOOK?

Last, I would suggest that science is not just a way to do things and that it is not just a body of knowledge, although it is indeed both of those things. I submit that it is also a way of looking at the world. The scientific outlook, the scientific worldview, offers humanity answers to central questions that humanity has always asked, although most people tend not to like the answers—that we are alone in the universe, that there is no ultimate meaning to life, but that people can be good and do good and offer the world things with science that religion and belief in God and the supernatural does not offer humanity. Certainly science has done far more to benefit humanity than religion has. Religious people can be very cruel. As Steven Weinberg says in that marvelous line of his, there will always be good people and there will always be bad people, but it takes religion to turn a good person into a bad person.

Religious belief motivated men to kill three thousand innocents on 9/11; these were the guys who believed God was on their side. And though I am not very well versed in the Bible, I do know that it includes many horrid stories where people are killing people, even instructed to do so by God. Children should simply not be exposed to that kind of negative influence. Of course, it doesn't always happen that way. There are both good and bad religious people, and there are both good and bad atheists.

THE SCIENTIFIC ALTERNATIVE

So what does science offer humanity that religion does not offer? The good life. All of the good things we appreciate in our world today are results of science and the applications of science in new technologies. The good things we experience on a daily basis in our society are not *in any way* the results of religion. Did religion bring us aspirin or air travel, heart surgery, or an increase of life expectancy? Did religion bring us computers and the Internet, modern drugs that heal the sick, or the ability to instantly communicate with one another even when separated by continents? Science has given us the good life; religion has detracted from it.

Some have even argued that life without God is more meaningful and good than life with religion. I am not entirely prepared to say that. I don't know if it's more or less fulfilling, but I do know that it is the only thing that makes sense. Of course, there are some people who are not rational and, therefore, don't see a problem with religious belief. They say, "I don't care if believing in God doesn't make sense, it works for me, it makes me feel good." Others use bad arguments to convince themselves that believing in God does make sense. I have become absolutely convinced that there is no argument you can make to such a person; there is no way to talk to such a person.

Science offers humanity something else that religion does not, and that's the utter beauty of science. The beauty and the wonder of science can fill your chest, and it is based on reality, not falsehoods. I concede that not everyone sees and appreciates the beauty of science, in fact, most people don't. I look at the Platonic solids and find them beautiful, or a mathematical proof to be absolutely beautiful; others see the material world and see its beauty as evidence of an immaterial world, which dumbfounds me.

Should it be a task of educators to try and broaden that group of people who can appreciate and find science beautiful, even with its implications for religious belief? It would be good if this was something that could be taught, but I'm somewhat skeptical that it can even be taught. Again, I believe people are the way they are. People tend to be religious and to not appreciate science. Exceptions are rare and result not from formal education entirely, I think, but from a kind of quality of the individual. There are many branches of mathematics that are absolutely beautiful, but most people live their whole lives and never hear of any proof of quadratic reciprocity, for instance. Or Gauss's theorem. Or Lagrange's theorems in number theory or in group theory. Some of the most beautiful theorems in mathematics are learned by and known to only 200 or 250 people each year.

Carl Gauss proved his theorem on quadratic reciprocity in probably 1810 or 1820. He loved that theorem so much, and he did no fewer than I think ten or eleven proofs. Now this is a major theorem of such beauty and such fundamental importance that by now there must be at least a couple hundred proofs around the world. Now, unless you know a little bit about the theory of numbers, it would not even be possible to communicate that beauty to someone, to even let someone begin to approach the beauty of it, so that they can appreciate it. It would be like showing a chimpanzee a Cézanne.

ON THE BEAUTY OF SCIENCE

I am aware of certain movements within the field of education that seem to popularize and foster the appreciation of science, and the work of such individuals: Richard Dawkins at Oxford, who is I think the preeminent example, or Carl Sagan and his work with science popularization, including his documentary series *Cosmos*, which I actually have never viewed, although I am aware it is the most widely watched documentary series in public television history. Some science educators think it's the chief mission of scientists and especially science educators to move in the direction of science appreciation, not just science education, similar to music appreciation courses. On the whole, I think it's a good idea because it would mean giving new sources of pleasure to large numbers of people. It would also reduce the rising tide of religious belief, I think.

If I didn't have the kind of environment I had as a child, I would have missed out on all of this beauty and pleasure.

Chapter 5

X-RAY CRYSTALLOGRAPHY

A History of Ideas

1. INTRODUCTION

In this account it is my aim to write about some of the ideas which have made possible the science of x-ray crystallography as we know it today. X-ray crystallography, since its birth in 1912, has undergone an explosive development. This rapid growth is no doubt due in part to the fact that this science lies at the intersection of many scientific disciplines: chemistry, physics, mathematics, materials science, biology, and the other life sciences. Not only have these sciences benefited from the rapid development of crystallography in the twentieth century but the phenomenal growth of the science of x-ray crystallography in turn was made possible through its interactions with these diverse scientific disciplines.

A remarkable feature of the ideas and discoveries which made possible the development of x-ray crystallography in the twentieth century is that they were, in large part, conceived many years, in some cases centuries, before the birth of x-ray crystallography itself. Thus these ideas, at the time they were formulated, had no obvious relationship to crystallography; they had been conceived instead for some other purpose having no apparent connection to our science. What were these ideas and what were their connections with x-ray crystallography?

A. Domenicano and I. Hargittai, eds., *Strength from Weakness: Structural Consequences of Weak Interactions in Molecular, Supermolecules, and Crystals.*

My intention in writing the first half of this chapter is to give some of the answers to these questions and to stress that, in large part, these ideas and discoveries could not, or would not, have been made had not the previous ones been made first.

The second half of this chapter is, on the other hand, dedicated to those ideas having a direct and obvious relationship to x-ray crystallography. They were conceived with the specific purpose of solving some problems in x-ray crystallography itself.

2. THE CALCULUS—NEWTON

Probably the invention of the calculus by Sir Isaac Newton and Leibniz is as good a starting point as any. It almost goes without saying that the calculus is an essential prerequisite not only for the science of x-ray crystallography but, probably without exception, for every one of the natural sciences as well. Although the motivation for Newton's invention is not known with certainty, it is known that one of the earliest applications was to "explain," in some sense, Kepler's three laws of planetary motion. To reach this goal Newton had first to formulate his laws of motion and the inverse square law of universal gravitation. Once this was done the calculus provided the tool which enabled Newton to obtain, for the first time, a satisfying description of the laws of planetary motion—a challenge which had defeated the best efforts of astronomers for centuries.

With respect to x-ray crystallography it suffices to refer, for example, to the phase problem—for decades the central challenge for theoretical crystallographers. Because of the strong mathematical flavor of this problem, it should not come as a surprise that the calculus would have a major role to play in its solution. For example, advanced techniques of the differential and integral calculus were required to perform the multiple integrations as well as the simplifying series approximations needed before significant progress could be made.

3. LEAST SQUARES—GAUSS

At the beginning of the nineteenth century astronomers were faced with the problem of tracking the newly discovered asteroids. Because of their small size, they could be seen with the telescopes available at that time for only short periods of time, insufficient to calculate an accurate orbit. The result was that these minor planets could not be reliably identified when they completed their orbits and returned to the vicinity of the Earth. The resulting confusion made it impossible to properly enumerate them.

Carl Friedrich Gauss, certainly the leading mathematician of the nineteenth century and possibly of all time, pointed the way. Very likely motivated by the astronomers' own problem with the asteroids, Gauss formulated his principle of least squares which, very much like the calculus, has turned out to be absolutely indispensable in facilitating progress in all the natural sciences, including the science of x-ray crystallography. For the astronomers' problem of the early nineteenth century it enabled Gauss to calculate with unprecedented precision the orbit if the asteroid Ceres despite the sparsity and the inherent experimental errors of the available observational data. Thus the astronomers' immediate problem was solved by Gauss, a teenager at the time.

With respect to x-ray crystallography it is of course common knowledge that the principle of least squares is almost universally used for the refinement of crystal structures. Least squares, as is well known, enables the crystallographer to determine, in a well-defined sense, the best crystal structure together with well-defined measures of standard deviation, consistent with the available, redundant set of experimentally observed diffraction intensities which are themselves subject to unavoidable errors of observation.

4. GROUP THEORY—GALOIS

Although no single mathematician can be credited with the initiation of the theory of groups, probably the young French mathematician Évariste Galois played as important a role as any. By the beginning of the nineteenth century the mathematicians had succeeded in solving the general equations of degrees one, two, three, and four using only the rational operations (addition, subtraction, multiplication, and division) and the extraction of roots. However, the solution of the general quintic defeated all their efforts. It remained for Galois, employing groups of substitutions operating on the roots of the equation, to find the necessary and sufficient condition that any equation be solvable by means of radicals and to show, in passing, that the general quintic could not be so solved.

Who could have anticipated, some two hundred years ago, that the theory of groups, devised by the mathematicians to solve algebraic equations, would turn out to be the perfect instrument for the study of crystallographic symmetry? For the crystallographer, however, the relevant groups were transformation groups, not groups of substitutions. However, the mathematicians, in their quest for generality, had developed the theory of abstract groups which included, as a special case, the 230 space groups of importance to crystallographers. Thus, once again, a concept devised for the solution of one problem turned out to be the indispensable tool for the solution of another, having no apparent relationship to the first.

5. HARMONIC ANALYSIS—FOURIER

About the beginning of the nineteenth century the French mathematician-physicist Jean Baptiste Fourier, in formulating his theory of heat, was confronted with problems in heat conduction and associated boundary value problems. Fourier based his solutions of these problems on his study of trigonometric series—series whose terms are the trigonometric functions sin nx and cos nx, n = 0, 1, 2, . . . , and which are known today as the Fourier series.

As is well known the trigonometric functions sin nx and cos nx are periodic functions of x with period 2π. Hence a Fourier series, under certain well-known conditions, represents a periodic function having the period 2π. Conversely, again under certain well-defined conditions, a function $f(x)$ of period 2π admits a Fourier series representation with Fourier coefficients determined in a known way by $f(x)$.

Owing to the three-dimensional periodic nature of crystals, the function r(r) which represents the electron density function in a crystal is a triply periodic function of the position vector r. Hence, again under conditions usually satisfied by the electron density function, r(r) admits a representation by a three-dimensional Fourier series with coefficients expressible in terms of r(r). It turns out that these coefficients are closely related to the diffraction intensities so that the well-established properties of Fourier series are indispensable to x-ray crystallographic analysis.

Here again we find that the study of Fourier series, initiated by Fourier in developing his theory of heat and having no obvious connection with crystallography, nevertheless has the most important consequences for x-ray crystallography.

6. X-RAYS—RÖNTGEN

In 1895 Wilhelm Röntgen was studying the properties of electrons flowing through a glass tube. He observed that whenever the flow of electrons was initiated a barium platinocyanide screen at some distance from the tube showed a flash of fluorescence. After some study he concluded that the fluorescence was due to some mysterious ray, generated by the impact of the electron beam on the glass wall of the tube and traveling in a straight line to the fluorescent screen. The discovery of x-rays by Röntgen in 1895 made possible the birth of x-ray crystallography seventeen years later.

7. EWALD'S THESIS—LAUE'S INTUITION

7.1. Introduction

Friedrich, Knipping, and Laue's discovery of the diffraction of x-rays by crystals in 1912 was a watershed event in modern science, and for this discovery Max van Laue won the 1914 Nobel Prize in Physics. When the very first Nobel Prizes had been awarded in 1901, the physics prize had gone to Wilhelm Röntgen for the discovery of x-rays, and in the years since 1912, right up to the present time, there has been a succession of Nobel Prizes awarded for work in or involving x-ray physics, spectroscopy, diffraction, and crystallography.

7.2. The Discovery of X-Ray Diffraction

In 1910 Paul Ewald[1] started work on his doctoral thesis problem, "To find the optical properties of an anisotropic arrangement of isotropic resonators," at Sommerfeld's Institute for Theoretical Physics in Munich. When Laue learned of Ewald's calculations in 1912, he asked whether Ewald's work was valid also for wavelengths smaller than the distance between neighboring resonators. When Ewald answered that it was, it occurred to Laue that a crystal, the atoms of which are arranged in a regular array, would diffract an incident beam of x-rays, with wavelengths comparable to the distances between neighboring atoms of the crystal, in accordance with Ewald's results. Laue then urged Friedrich, an assistant to Sommerfeld, and Knipping, who had just completed his thesis with Röntgen, to perform the experiment. The results showed that crystals do indeed act as a three-dimensional diffraction grating for x-rays, thus dramatically confirming Laue's insight. Within weeks Laue had worked out the mathematical description of the diffraction phenomenon and correctly interpreted the features of the diffraction pattern.

8. DIFFRACTION AS SPECULAR REFLECTION—THE BRAGGS—THE FIRST CRYSTAL STRUCTURE DETERMINATIONS

Within the year, W. H. and W. L. Bragg,[2] father and son, simplified Laue's mathematical description of the diffraction conditions by introducing the idea of specular reflection from the atomic planes within the crystal. They deduced the celebrated Bragg equation

$$nl = 2d \sin q$$

where n is an integer called the order of the reflection, l is the wavelength of the x-rays, and d is the repeat spacing between the atomic planes for which q is the common angle of incidence and reflection. Through this relationship they were able to measure interatomic distances, make structural chemistry quantitative, and thus establish the science of x-ray crystallography; they were able also to measure the x-ray wavelengths and thus establish the science of x-ray spectroscopy. With this background the Braggs could then relate diffraction patterns with crystal structures and determine the atomic arrangements in crystals of simple substances such as NaCl, KCl, KBr, and KI. They could not possibly have anticipated in those early years that the method they pioneered would, by the 1970s, have been so strengthened that even proteins and nucleic acids, having thousands of atoms in the molecule, would yield their structures to this powerful technique. Even the secrets of virus structures, many times more complex still, are now being revealed.

9. THE PHASE PROBLEM

After the discovery of the diffraction of x-rays by crystals by Friedrich, Knipping, and Laue and the determination of the simplest crystal structures by the Braggs, it was generally understood that the key to the determination of molecular structures, that is, the geometric arrangement of the atoms which constitute a molecule, had finally been found. Of course, in those early years it was natural to use a trial and error technique in the actual structure determination. What this required was first to postulate a plausible structure based usually on crystal habit, prior chemical knowledge, and intuition. On the basis of such an informed guess, one could calculate the nature of the diffraction pattern, that is, the directions and relative intensities of the scattered x-rays. By comparison of the calculated with the observed diffraction patterns one could then confirm or reject the assumed structure. If the two diffraction patterns were in good agreement, one could safely assume that the postulated structure was essentially correct, and standard iterative techniques were then employed to refine the initial guess. For some years this primitive technique was used to solve a number of simple structures. However, it soon became clear that more powerful techniques would be required to solve the more complex molecular structures of interest to chemists, mineralogists, and biologists.

One of the consequences of the early experiments was the recognition that x-rays, like ordinary visible light and radio waves, were an electromagnetic disturbance. Hence x-rays have a frequency, a wavelength, and a phase, as well as intensity. Furthermore, in interpreting the x-ray scattering experiment as a diffraction effect one could calculate, on the basis of an assumed crystal structure, not only the directions and intensities of the scattered rays but their phases as well. Finally, and this is a matter of the greatest importance, it was also shown in those early days that, conversely, from a complete knowledge of the diffraction pattern, that is, the directions, intensities, and phases of the scattered rays, one could deduce unambiguously the crystal structure, that is, the electron density function and therefore the precise arrangement of the atoms in the crystal. Unfortunately, the phases of the scattered rays could not be measured; they were lost in the diffraction experiment. Since one could therefore use arbitrary values for the lost phases together with the measured intensities, many crystal structures, all presumably consistent with the observed intensities, could be obtained. This argument led the chemists and crystallographers of those early years to the pessimistic conclusion that the diffraction experiment could not, after all, lead unambiguously to unique crystal structures, even in principle.

10. PATTERSON'S IDEA

If measured intensities could not, even in principle, lead to a unique crystal structure, it was natural to ask what do the intensities determine? It was A. L. Patterson[3] who answered this question in 1934 when he showed that the information content of the measured intensities coincides precisely with the collection of all the interatomic vectors in the crystal. If there are N atoms in the unit cell of the crystal there are N^2 interatomic vectors, so that the difficulty in solving the problem of going from a knowledge of all the interatomic vectors to a knowledge of the crystal structure increases rapidly with increasing complexity of the crystal structure. In practice, then, Patterson's method has proven to be useful only for simple structures or for more complex structures containing one or a small number of "heavy" atoms, that is, atoms with high atomic numbers. In such cases the Patterson synthesis is capable of yielding the positions of the heavy atoms from which the remainder of the structure can often be deduced. For complex structures there are so many interatomic vectors, many of which usually coincide, that the Patterson function is poorly resolved, the interatomic vectors cannot be well determined, and the further analysis leading to the structure is usually precluded. Even if one exploits the crystallographic symmetry, as David Harker proposed to do in 1936, thus greatly increasing the power of the technique, the method has limited usefulness for complex structures unless a small number of heavy atoms is present. Thus by 1950 the problem of determining complex molecular structures having no heavy atoms, the so-called all light atom structures, which were often of greatest interest to chemists and biologists, was generally regarded as insoluble.

11. THE PHASE PROBLEM AGAIN

For almost forty years, then, following the discovery of the diffraction of x-rays by crystals in 1912, it was generally believed, for reasons already given, that the diffraction intensities alone were insufficient to determine crystal structure unambiguously. This belief was finally refuted in the early 1950s by the recognition that a priori structural knowledge, when combined with the measurable diffraction intensities, did, in fact, provide sufficient information to lead, in general, to unique crystal and molecular structures. The methods devised to achieve this goal are known as direct methods. These methods show that, because real crystal structures must satisfy certain restrictive conditions, relationships exist among the intensities and phases of the scattered x-rays which permit the phases to be recovered once the intensities are known. Thus the phase information, which is lost in the diffraction experiment, is, in fact, to be found among the measurable intensities. In short, the phase problem, which is to determine the values of the missing phases from the known diffraction intensities, is a solvable one, and the solution technique is called the direct method.

12. PHASE RELATIONSHIPS

The overdetermination of the phase problem implies that a large number of identities among the phases must, of necessity, exist. Only two of the most important of these, the Sayre equation and the tangent formula, will be briefly mentioned here.

12.1. The Sayre Equation

In 1952 David Sayre,[4] using his "squaring method," which exploits the connection between a crystal structure and its square, derived the celebrated Sayre equation, an explicit relationship among the phases, dependent of course on measured diffraction intensities. This equation has withstood the test of time so that, even today, it finds application in many computer programs for the direct determination of phases.

12.2. The Tangent Formula

Closely related to the Sayre equation is the tangent formula, first derived by Karle and Hauptman[5] in 1956 using probabilistic techniques. The tangent formula, in one form on another, is almost universally used in computer programs for direct phase determination. It represents one of the earliest examples of the probabilistic approach to the phase problem and serves to illustrate the power of probabilistic methods on which the direct methods of phase determination are primarily based.[6]

13. NON-NEGATIVITY AND ATOMICITY

What is the nature of this a priori structural knowledge? It is of two kinds. The first is simply that the electron density function in a crystal is nonnegative everywhere. This property of crystal structures leads to inequality relationships among the known intensities and the desired phases. The first to discover some of these relationships were Harker and Kasper,[7] who published their work in 1948. Implicit in their derivation was the assumption that the density function is nonnegative. For this reason their celebrated paper must be regarded as the first in the long series of papers on direct methods which followed.

In 1950 Karle and Hauptman[8] published their paper on inequalities in which the logical basis, that is, the nonnegativity of the electron density function, of the Harker-Kasper inequalities was stressed. Furthermore, all inequalities based on the nonnegativity property were derived.

Although inequality relationships severely restrict the values of the phases, they appear to be insufficient to determine unique values for the phases. In addition, because of their complexity, they are difficult to apply in practice, so that they have not proved to be useful except for very simple structures. A much more restrictive and useful property of crystal structures is atomicity, and it is this property which constitutes the foundation on which the structure of direct methods is based.

Molecules consist of atoms. Hence the electron density function in a crystal is not only nonnegative everywhere but it reaches maximum positive values at the positions of the atoms and drops down to small values between the atomic positions. This property of crystal structures, together with the large number of diffraction intensities available from experiment, is, in general, sufficiently restrictive to determine unique values for the phases of the scattered x-rays. It turns out that the measured intensities are more than sufficient for this purpose so that the phase problem is not only a solvable one, at least in principle, but is actually a greatly overdetermined one.

14. PROBABILISTIC METHODS

Who would have thought that the theory of mathematical probability would find application in the solution of the phase problem? Here is how that has come about. If one replaces the atomic position vectors r of the (point) atoms in a crystal by random variables, uniformly and independently distributed, the (complex valued) normalized structure factors E, as functions of the atomic position vectors r, are themselves random variables having probability distributions which may be found by standard methods of mathematical probability. It is then a straightforward matter to derive the conditional probability distributions of arbitrary combinations of the phases assuming that the magnitudes $|E|$, or what is the same thing, the diffraction intensities, are known. In this way relationships among the phases, having probabilistic validity, are readily found. One is then led to the solution of the phase problem by combining the information contained in these relationships. This is best done by formulating the phase problem as a problem in constrained global minimization.

15. THE MINIMAL PRINCIPLE— THE PHASE PROBLEM AS A PROBLEM IN CONSTRAINED GLOBAL MINIMIZATION

The probability distributions described in the previous section lead to the definition of the minimal function $m(\mathrm{j})$, a function of the phases j dependent on the values of the magnitudes $|E|$, presumed to be known from the observed diffraction intensities. It then follows from the probabilistic background that the correct phases are those for which the minimal function $m(\mathrm{j})$ reaches its constrained global minimum (the minimal principle). The constraints arise because, as pointed out earlier, the overdetermination of the phase problem implies that identities among the phases must of necessity be satisfied.

16. THE COMPUTER PROGRAM *SHAKE-AND-BAKE*

It is one thing to formulate the phase problem as a problem in constrained global minimization. It is quite another to devise a technique for finding the constrained global minimum of the minimal function $m(\mathrm{j})$. The first algorithm to do this, called *Shake-and-Bake*,[9] by alternating phase refinement in reciprocal space (via the minimal function) with density modification in real space, provides a routine, ab initio, and automatic solution to the phase problem. Variations of this algorithm, based on similar principles (e.g., half-baked, SHELX, etc.) have since been devised. These algorithms have solved routinely structures having as many as two thousand atoms in the unit cell. They all require diffraction intensities to atomic resolution, about 1.2 Å.

It should be stressed, in conclusion, that the methods of mathematical probability, having no apparent relationship to the phase problem of x-ray crystallography, nevertheless are instrumental in providing the most powerful methods for solving this all-important problem.

NOTES

1. P. P. Ewald, ed., *Fifty Years of X-ray Diffraction* (Utrecht: Oosthoek's, 1962).

2. L. Bragg, *The Development of X-ray Analysis* (London: Bell and Sons, 1975).

3. J. P. Glusker, ed., *Structural Crystallography in Chemistry and Biology*, vol. 4 of *Benchmark Papers in Physical Chemistry and Chemical Physics* (New York: Hutchinson Ross, 1981).

4. D. Sayre, "The Squaring Method: A New Method for Phase Determination," *Acta Crystallographica* 5 (1952): 60–65.

5. J. Karle and H. Hauptman, "A Theory of Phase Determination for the Four Types of Non-centrosymmetric Space Groups $1P222$, $2P22$, $3P_12$, $3P_22$," *Acta Crystallographica* 9 (1956): 635–51.

6. D. McLachlan and J. P. Glusker, eds., *Crystallography in North America* (New York: American Crystallographic Association, 1983).

7. D. Harker and J. S. Kasper, "Phases of Fourier Coefficients Directly from Crystal Diffraction Data," *Acta Crystallographica* 1 (1948): 70–75.

8. J. Karle and H. Hauptman, "The Phases and Magnitudes of the Structure Factors," *Acta Crystallographica* 3 (1950): 181–87.

9. G. M. Sheldrick, H. A. Hauptman, C. M. Weeks, R. Miller, and J. Usón, "Ab Initio Phasing," in *International Tables for Crystallography*, ed. M. G. Rossmann and E. Arnold, vol. F of *Crystallography of Biological Macromolecules* (Dordrecht: Kluwer, 2001), pp. 333–51.

APPENDICES

Appendix A

NOBEL PRIZE PRESENTATION SPEECH

Presentation Speech by Professor Ingvar Lindqvist
of the Royal Academy of Sciences

Translation from the Swedish Text

Your Majesties, Your Royal Highnesses, Ladies and Gentlemen,

The youth of today find it quite natural that there are such things as atoms and molecules. They have often seen molecular models in school and experienced the molecule as something obviously existing. There have been people over thousands of years who have come to the conclusion—instinctively, emotionally, or logically—that molecules should exist. These people have also had ideas about the shape and properties of the molecules.

These efforts reached their climax at the end of the nineteenth century in three extraordinary theories: the idea by van't Hoff about the significance of the tetrahedral carbon atom, the revelation by Kekulé of the structure of benzene, and the description by Werner of many metal complexes as having octahedral, tetrahedral, or planar square structures. The geniality of these ideas has been strongly confirmed in our century to an extent, which proves how epoch-making these achievements were.

From *Nobel Lectures, Chemistry 1981–1990*, editor-in-charge, Tore Frängsmyr; ed. Bo G. Malmström (Singapore; River Edge, NJ: World Scientific, 1992).

APPENDIX A

It is not, however, until the twentieth century that the scientists have created methods which admit a complete determination of the structures of molecules. In this context structure means the geometrical arrangement of atoms as well as the bonding distances between atoms. The most important of these methods is x-ray crystallography.

In such investigations the x-rays are arranged to strike a crystal. The radiation is then scattered in certain directions and the light intensity is measured for each such scattered x-ray. Such an experiment was first made by Max von Laue, who obtained the Nobel Prize for Physics in 1914 for his discovery. The Braggs, father and son, made the first structure determinations of simple chemical compounds and were awarded the Nobel Prize for Physics in 1915.

It was not, however, possible to determine crystal structures without some assumptions or guesses, because the phase differences between the different scattered x-rays were not known. The crystallographers had to use a trial-and-error method.

Several methodological improvements have since taken place, but it has for a long time been considered a great scientific achievement to determine the molecular structures of organic molecules as large as penicillin or vitamin B12. As late as 1964, Dorothy Hodgkin was awarded the Nobel Prize in Chemistry for such structure determinations.

It therefore was met with great interest and much opposition and discussion when Herbert Hauptman and Jerome Karle during the years 1950–1956 published a series of papers in which they claimed to have found a general method, a "direct" method for solving the phase problem, thus opening the possibility to determine the structure directly from the experimental results without any further assumptions. Hauptman and Karle built their method on two established facts. One was that the electron density in a molecule can never be negative—there are electrons or there are not. The other fact was that the number of experimental results is large enough to permit application of statistical methods. Recent developments have shown that they were right and the production of modern computers has strongly contributed to the rapidity and efficiency of their methods. These methods are now so efficient that structure determinations for which the Nobel Prize was awarded in 1964 can today be made by a clever beginner.

At the same time it has been more and more important for the chemists to know the exact structures of molecules which take part in important chemical and biochemical reactions. One could without exaggeration say that it is only in the last ten years that chemistry has developed into a truly molecular era. Molecules with desired structures and properties can be produced and the molecular mechanism is known for increasingly more reactions.

It is this importance to chemistry which has motivated a Nobel Prize in Chemistry to the mathematician Herbert Hauptman and the physicist Jerome Karle. Another way to express it is that the imagination and ingenuity of the laureates have made it unnecessary to exercise these qualities in normal structure determinations. On the other hand, they have increased the possibilities for the chemists to use their imagination and their ingenuity.

Herbert Hauptman and Jerome Karle,

Your basic development of the direct methods for x-ray crystallographic structure determination has given the chemists an efficient tool for faster and more detailed studies of the structures of molecules and therefore also for the study of chemical reactions. On behalf of the academy, I wish to convey to you our warmest congratulations and I now ask you to receive your prizes from the hands of His Majesty the King.

Appendix B

NOBEL PRIZE ACCEPTANCE SPEECH

Herbert A. Hauptman's Speech at the Nobel Banquet,
December 10, 1985

Your Majesties, Your Royal Highnesses, Ladies and Gentlemen,

I speak for Jerome Karle, as well as myself, when I say that our journey to Stockholm began some sixty-seven years ago when our parents, with unconscious wisdom, gave us a most precious gift, the freedom to grow as we wished, at our own pace, and in the direction of our own choosing. We chose to read a great deal, as soon as we were able, in all areas of science. To their credit our parents permitted, even encouraged, this activity when there may have been moments when they secretly questioned the wisdom of our course. We wish, on this occasion, to make grateful acknowledgment of our indebtedness to them for their sacrifices on our behalf.

We are grateful, too, for the opportunity to have attended the City College of New York, at a time when a free education was provided to those who qualified and who would not otherwise have been able to obtain a higher education. Without this splendid gift, it is doubtful that we would be here today.

We are also indebted to the Naval Research Laboratory for supporting us in our pursuit of scientific knowledge for its own sake.

From *Les Prix Nobel. The Nobel Prizes 1985*, ed. Wilhelm Odelberg (Nobel Foundation: Stockholm, 1986).

APPENDIX B

We wish finally to thank our wives for their continuing support and encouragement, particularly during the early years when our work was received with some skepticism.

We were fortunate, too, that our particular qualifications, Jerome Karle's in physical chemistry and mine in mathematics, were the exact combination which was needed to enable us to tackle, with some hope of success, the phase problem of x-ray crystallography, the major stumbling block in the solution of crystal structures by the technique of x-ray diffraction. Our sole motivation was to overcome the challenge which this problem presented, and our satisfactions came from the progress we made. We were fortunate, indeed, that the implications for structural chemistry turned out to be so far reaching; we did not anticipate them.

In summary then, we are grateful to all, including our scientific colleagues, who by their support helped bring us to this place; and we also readily acknowledge that good fortune played a major role.

Appendix C

NEW YORK TIMES ARTICLE

Scientists Speak Up on Mix of God and Science

Cornelia Dean

August 23, 2005

At a recent scientific conference at City College of New York, a student in the audience rose to ask the panelists an unexpected question: "Can you be a good scientist and believe in God?"

Reaction from one of the panelists, all Nobel laureates, was quick and sharp. "No!" declared Herbert A. Hauptman, who shared the chemistry prize in 1985 for his work on the structure of crystals.

Belief in the supernatural, especially belief in God, is not only incompatible with good science, Dr. Hauptman declared, "this kind of belief is damaging to the well-being of the human race."

But disdain for religion is far from universal among scientists. And today, as religious groups challenge scientists in arenas as various as evolution in the classroom, AIDS prevention and stem cell research, scientists who embrace religion are beginning to speak out about their faith.

"It should not be a taboo subject, but frankly it often is in scientific cir-

---

"Scientists Speak Up on Mix of God and Science" from the *New York Times*, national section, August 23, 2007, issue, A1. Reprinted with permission.

cles," said Francis S. Collins, who directs the National Human Genome Research Institute and who speaks freely about his Christian faith.

Although they embrace religious faith, these scientists also embrace science as it has been defined for centuries. That is, they look to the natural world for explanations of what happens in the natural world and they recognize that scientific ideas must be provisional—capable of being overturned by evidence from experimentation and observation. This belief in science sets them apart from those who endorse creationism or its doctrinal cousin, intelligent design, both of which depend on the existence of a supernatural force.

Their belief in God challenges scientists who regard religious belief as little more than magical thinking, as some do. Their faith also challenges believers who denounce science as a godless enterprise and scientists as secular elitists contemptuous of God-fearing people.

Some scientists say simply that science and religion are two separate realms, "non-overlapping magisteria," as the late evolutionary biologist Stephen Jay Gould put it in his book *Rocks of Ages* (Ballantine, 1999). In Dr. Gould's view, science speaks with authority in the realm of "what the universe is made of (fact) and why does it work this way (theory)" and religion holds sway over "questions of ultimate meaning and moral value."

When the American Association for the Advancement of Science devoted a session to this idea of separation at its annual meeting this year, scores of scientists crowded into a room to hear it.

Some of them said they were unsatisfied with the idea, because they believe scientists' moral values must inevitably affect their work, others because so much of science has so many ethical implications in the real world.

One panelist, Dr. Noah Efron of Bar-Ilan University in Israel, said scientists, like other people, were guided by their own human purposes, meaning and values. The idea that fact can be separated from values and meaning "jibes poorly with what we know of the history of science," Dr. Efron said.

Dr. Collins, who is working on a book about his religious faith, also believes that people should not have to keep religious beliefs and scientific theories strictly separate. "I don't find it very satisfactory and I don't find it very necessary," he said in an interview. He noted that until relatively recently, most scientists were believers. "Isaac Newton wrote a lot more about the Bible than the laws of nature," he said.

But he acknowledged that as head of the American government's efforts to decipher the human genetic code, he had a leading role in work that many say definitively demonstrates the strength of evolutionary theory to explain the complexity and abundance of life.

As scientists compare human genes with those of other mammals, tiny worms, even bacteria, the similarities "are absolutely compelling," Dr. Collins said. "If Darwin had tried to imagine a way to prove his theory, he could not have come up with something better, except maybe a time machine. Asking somebody to reject all of that in order to prove that they really do love God—what a horrible choice."

Dr. Collins was a nonbeliever until he was 27—"more and more into the mode of being not only agnostic but being an atheist," as he put it. All that changed after he completed his doctorate in physics and was at work on his medical degree, when he was among those treating a woman dying of heart disease. "She was very clear about her faith and she looked me square in the eye and she said, 'What do you believe?'" he recalled. "I sort of stammered out, 'I am not sure.'"

He said he realized then that he had never considered the matter seriously, the way a scientist should. He began reading about various religious beliefs, which only confused him. Finally, a Methodist minister gave him a book, *Mere Christianity*, by C. S. Lewis. In the book Lewis, an atheist until he was a grown man, argues that the idea of right and wrong is universal among people, a moral law they "did not make, and cannot quite forget even when they try." This universal feeling, he said, is evidence for the plausibility of God.

When he read the book, Dr. Collins said, "I thought, my gosh, this guy is me."

Today, Dr. Collins said, he does not embrace any particular denomination, but he is a Christian. Colleagues sometimes express surprise at his faith, he said. "They'll say, 'How can you believe that? Did you check your brain at the door?'" But he said he had discovered in talking to students and colleagues that "there is a great deal of interest in this topic."

APPENDIX C

POLLING SCIENTISTS ON BELIEFS

According to a much-discussed survey reported in the journal *Nature* in 1997, 40 percent of biologists, physicists and mathematicians said they believed in God—and not just a nonspecific transcendental presence but, as the survey put it, a God to whom one may pray "in expectation of receiving an answer."

The survey, by Edward J. Larson of the University of Georgia, was intended to replicate one conducted in 1914, and the results were virtually unchanged. In both cases, participants were drawn from a directory of American scientists.

Others play down those results. They note that when Dr. Larson put part of the same survey to "leading scientists"—in this case, members of the National Academy of Sciences, perhaps the nation's most eminent scientific organization—fewer than 10 percent professed belief in a personal God or human immortality.

This response is not surprising to researchers like Steven Weinberg, a physicist at the University of Texas, a member of the academy and a winner of the Nobel Prize in 1979 for his work in particle physics. He said he could understand why religious people would believe that anything that eroded belief was destructive. But he added: "I think one of the great historical contributions of science is to weaken the hold of religion. That's a good thing."

NO GOD, NO MORAL COMPASS?

He rejects the idea that scientists who reject religion are arrogant. "We know how many mistakes we've made," Dr. Weinberg said. And he is angered by assertions that people without religious faith are without a moral compass.

In any event, he added, "The experience of being a scientist makes religion seem fairly irrelevant." He said, "Most scientists I know simply don't think about it very much. They don't think about religion enough to qualify as practicing atheists."

Most scientists he knows who do believe in God, he added, believe in "a God who is behind the laws of nature but who is not intervening."

Kenneth R. Miller, a biology professor at Brown, said his students were often surprised to find that he was religious, especially when they realized that his faith was not some sort of vague theism but observant Roman Catholicism.

Dr. Miller, whose book, *Finding Darwin's God*, explains his reconciliation of the theory of evolution with his religious faith, said he was usually challenged in his biology classes by one or two students whose religions did not accept evolution, who asked how important the theory would be in the course.

"What they are really asking me is 'do I have to believe in this stuff to get an A?'" he said. He says he tells them that "belief is never an issue in science."

"I don't care if you believe in the Krebs cycle," he said, referring to the process by which energy is utilized in the cell. "I just want you to know what it is and how it works. My feeling about evolution is the same thing."

For Dr. Miller and other scientists, research is not about belief. "Faith is one thing, what you believe from the heart," said Joseph E. Murray, who won the Nobel Prize in medicine in 1990 for his work in organ transplantation. But in scientific research, he said, "it's the results that count."

Dr. Murray, who describes himself as "a cradle Catholic" who has rarely missed weekly Mass and who prays every morning, said that when he was preparing for the first ever human organ transplant, a kidney that a young man had donated to his identical twin, he and his colleagues consulted a number of religious leaders about whether they were doing the right thing. "It seemed natural," he said.

USING EVERY TOOL

"When you are searching for truth you should use every possible avenue, including revelation," said Dr. Murray, who is a member of the Pontifical Academy, which advises the Vatican on scientific issues, and who described the influence of his faith on his work in his memoir, *Surgery of the Soul* (Science History Publications, 2002).

APPENDIX C

Since his appearance at the City College panel, when he was dismayed by the tepid reception received by his remarks on the incompatibility of good science and religious belief, Dr. Hauptman said he had been discussing the issue with colleagues in Buffalo, where he is president of the Hauptman-Woodward Medical Research Institute.

"I think almost without exception the people I have spoken to are scientists and they do believe in the existence of a supreme being," he said. "If you ask me to explain it—I cannot explain it at all."

But Richard Dawkins, an evolutionary theorist at Oxford, said that even scientists who were believers did not claim evidence for that belief. "The most they will claim is that there is no evidence against," Dr. Dawkins said, "which is pathetically weak. There is no evidence against all sorts of things, but we don't waste our time believing in them."

Dr. Collins said he believed that some scientists were unwilling to profess faith in public "because the assumption is if you are a scientist you don't have any need of action of the supernatural sort," or because of pride in the idea that science is the ultimate source of intellectual meaning.

But he said he believed that some scientists were simply unwilling to confront the big questions religion tried to answer. "You will never understand what it means to be a human being through naturalistic observation," he said. "You won't understand why you are here and what the meaning is. Science has no power to address these questions—and are they not the most important questions we ask ourselves?"

Appendix D

FREE INQUIRY INTERVIEW

At an August 2005 City College of New York conference featuring a panel of Nobel laureates, one scientist created a stir by arguing that belief in God is incompatible with being a good scientist and is "damaging to the well-being of the human race." Herbert Hauptman shared the Nobel Prize in Chemistry in 1985 for his work on the structure of crystals and is also a laureate of the International Academy of Humanism. A gentle, unassuming man in his eighties, Hauptman sat down with *Free Inquiry* at the acclaimed Hauptman-Woodward Institute in Buffalo, New York.

—D. J. Grothe, Associate Editor
Free Inquiry

FI: What led you to speak out about religion versus science?

Hauptman: City College is proud of its Nobel laureates, of which they have eight or nine, and we came to do a panel at a scientific conference and to serve as judges for contributions by CUNY students. After the panel, one of the students asked the question regarding the compatibility of science and religion. I ended up being the only one who answered the question, which surprised me.

FI: What response did you elicit from the audience?

Hauptman: There was little or no reaction . . . from the audience nor from the other panelists. I completely expected other panelists to support what I said, but none did. The only significant negative reaction came from Cornelia Dean, a reporter from the *New York Times*. I was later told by several of the other Nobel laureates that they agreed with me . . . but for reasons of their own, they just did not respond.

FI: Why do you think they were reticent?

Hauptman: Well, obviously this view is unpopular in this overly religious society. People who are outspoken about it are more than just regarded as cranky, they are deeply disliked.

FI: So why did you speak out?

Hauptman: I have never hidden my beliefs, but neither did I advertise them. In fact, I never thought too terribly much about it; I have kept myself busy thinking about other problems, scientific problems. But I spoke out because of this frustration I have only lately begun to feel about the religiosity in our society.

FI: Then came the media response. A story by Ms. Dean concerning your remarks appeared on the front page of the *New York Times*. Never having publicly aired your views on religion before, were you afraid of being thrust into the media as an atheist?

Hauptman: No, not really. I received a number of letters, mostly positive. But when a producer at *This Week with George Stephanopoulos* . . . invited me to appear on the show, my wife suggested I not do it out of concern for my safety. Consider the beating of the professor in Kansas who was attacked for announcing he was going to teach a course on evolution versus intelligent design. Or Bernard Slepian, the doctor who was slain for conducting abortions. Whenever you hear of these horrible acts of violence . . . you can be pretty sure they are not done because of someone's lack of belief in God, but out of a fervent religious belief. Of course, most religious fundamentalists are not violent. In any case, out of concern for my safety, we decided not to do Stephanopoulos.

FI: Over 90 percent of members of the prestigious National Academy of Sciences are atheists or agnostics. Do you think there is a relationship between being a good scientist and being a religious skeptic?

Hauptman: What are religions based on? They are not based on evidence, but on faith. On the other hand, a good scientist . . . insists that before one assents to a claim, there must be good evidence for that claim. . . . If you live by this principle of science, I believe you will end up believing as I and most of the other members of the National Academy of Sciences believe: that there is no God.

FI: What do you think of those scientists who believe as you do but refuse to let their views be known?

Hauptman: I do not think they should be in the closet on this issue, but it is really a matter of how you allocate your time and energy . . . and a matter of conscience. Still, I think we would be better off if scientists were more open about their lack of belief in God.

FI: What is one question about the science versus religion controversy that you would like answered?

Hauptman: When will religion no longer be an issue of importance to the majority of the people in our society?

Appendix E

DIRECT METHODS AND ANOMALOUS DISPERSION

Nobel Lecture

DIRECT METHODS AND ANOMALOUS DISPERSION

Nobel lecture, 9 December, 1985

by

HERBERT A. HAUPTMAN

Medical Foundation of Buffalo, Inc. 73 High Street, BUFFALO, N. Y. 14203

1. INTRODUCTION

The electron density function, $\varrho(\mathbf{r})$, in a crystal determines its diffraction pattern, i.e. both the magnitudes and phases of its x-ray diffraction maxima, and conversely. If, however, as is always the case, only magnitudes are available from the diffraction experiment, then the density function $\rho(\mathbf{r})$ cannot be recovered. If one invokes prior structural knowledge, usually that the crystal is composed of discrete atoms of known atomic numbers, then the observed magnitudes are, in general, sufficient to determine the positions of the atoms, i.e. the crystal structure.

It should be noted here that the recognition that observed diffraction data are in general sufficient to determine crystal structures uniquely was an important milestone in the development of the direct methods of crystal structure determination. The erroneous contrary view, that crystal structures could not, even in principle, be deduced from diffraction intensities, had long been held by the crystallographic community prior to c. 1950 and constituted a psychological barrier which first had to be removed before real progress could be made.

2. THE TRADITIONAL DIRECT METHODS

2.1. The phase problem.

Denote by $\phi_{\mathbf{H}}$ the phase of the structure factor $F_{\mathbf{H}}$:

$$F_{\mathbf{H}} = |F_{\mathbf{H}}| \exp(i\phi_{\mathbf{H}}), \tag{1}$$

where $\mathbf{H}$ is a reciprocal lattice vector (having three integer components) which labels the corresponding diffraction maximum. Then the relationship between the structure factors $F_{\mathbf{H}}$ and the electron density function $\varrho(\mathbf{r})$ is given by

$$F_{\mathbf{H}} = \int_V \rho(\mathbf{r}) \exp(2\pi i \mathbf{H}\cdot\mathbf{r}) dV \tag{2}$$

and

$$\rho(\mathbf{r}) = \frac{1}{V}\sum_{\mathbf{H}} F_{\mathbf{H}}\exp(-2\pi i\mathbf{H}\cdot\mathbf{r}) = \frac{1}{V}\sum_{\mathbf{H}}|F_{\mathbf{H}}|\exp i(\phi_{\mathbf{H}} - 2\pi\mathbf{H}\cdot\mathbf{r}) \tag{3}$$

in which V represents the unit cell or its volume. Thus the structure factors F_H determine $\rho(\mathbf{r})$. The x-ray diffraction experiment yields only the magnitudes $|F_H|$ of a finite number of structure factors, but the values of the phases $\phi_{\mathbf{H}}$, which are also needed if one is to determine $\rho(\mathbf{r})$ from (3), cannot be determined experimentally. If arbitrary values for the phases $\phi_{\mathbf{H}}$ are specified in Eq. (3), then density functions $\varrho(\mathbf{r})$ are defined which, when substituted into (2) yield structure factors F_H the magnitudes of which agree with the observed magnitudes $|\mathbf{F_H}|$. It follows that the diffraction experiment does not determine $\rho(\mathbf{r})$. It was this argument which led crystallographers, prior to 1950, to the erroneous conclusion that diffraction intensities could not, even in principle, determine crystal structures uniquely. What had been overlooked was the fact that the phases $\phi_{\mathbf{H}}$ could not be arbitrarily specified if (3) is to yield density functions characteristic of real crystals.

Crystals are composed of discrete atoms. One exploits this prior structural knowledge by replacing the real crystal, with continuous electron density $\rho(\mathbf{r})$, by an ideal one, the unit cell of which consists of N discrete, non-vibrating, point atoms located at the maxima of $\rho(\mathbf{r})$. Then the structure factor F_H is replaced by the normalized structure factor E_H and (1) to (3) are replaced by

$$E_{\mathbf{H}} = |E_{\mathbf{H}}| \exp(i\phi_{\mathbf{H}}), \tag{4}$$

$$E_{\mathbf{H}} = \frac{1}{\sigma_2^{1/2}} \sum_{j=1}^{N} f_j \exp(2\pi i \mathbf{H}\cdot\mathbf{r}_j), \tag{5}$$

$$\begin{aligned} \left\langle E_{\mathbf{H}} \exp(-2\pi i \mathbf{H}\cdot\mathbf{r})\right\rangle_{\mathbf{H}} &= \frac{1}{\sigma_2^{1/2}} \left\langle \sum_{j=1}^{N} f_j \exp\left[2\pi i \mathbf{H}\cdot(\mathbf{r}_j - \mathbf{r})\right] \right\rangle_{\mathbf{H}} \\ &= \frac{f_j}{\sigma_2^{1/2}} \text{ if } \mathbf{r} = \mathbf{r_j} \\ &= 0 \text{ if } \mathbf{r} \neq \mathbf{r_j} \end{aligned} \tag{6}$$

where f_j is the zero-angle atomic scattering factor, $\mathbf{r}_j$ is the position vector of the atom labelled j, and

$$\sigma_n = \sum_{j=1}^{N} f_j^n, \; n = 1, 2, 3, \ldots \tag{7}$$

In the x-ray diffraction case the f_j are equal to the atomic numbers Z_j and are presumed to be known. From (6) it follows that the normalized structure factors E_H determine the atomic position vectors $\mathbf{r_j}$, j = 1,2,..., N, i.e. the crystal structure.

In practice a finite number of magnitudes $|E_H|$ of normalized structure factors E_H are obtainable (at least approximately) from the observed magnitudes $|F_H|$ while the phases $\phi_{\mathbf{H}}$, as defined by (4) and (5), cannot be determined experimentally. Since one now requires only the 3N components of the N position vectors $\mathbf{r}_j$, rather than the much more complicated electron density function $\varrho(\mathbf{r})$, it turns out that, in general, the known magnitudes are more than sufficient. This is most readily seen by equating the magnitudes of both sides of (5) in order to obtain a system of equations in which the only unknowns are the 3N components of the position vectors $\mathbf{r}_j$. Since the number of such equations, equal to the number of reciprocal lattice vectors **H** for which magnitudes $|E_{\mathbf{H}}|$ are available, usually greatly exceeds the number, 3N, of unknowns, this system is redundant. Thus observed diffraction intensities usually over-determine the crystal structure, i.e. the positions of the atoms in the unit cell. In short, by merely replacing the integral of Eq. (2) by the summation of Eq. (5), i.e. taking Eq. (5) as the starting point of our investigation rather than Eq. (2), one has transformed the problem from an unsolvable one to one which is solvable, at least in principle.

In summary then, the intensities (or magnitudes $|E_{\mathbf{H}}|$) of a sufficient number of x-ray diffraction maxima determine a crystal structure. The available intensities usually exceed the number of parameters needed to describe the structure. From these intensities a set of numbers $|E_{\mathbf{H}}|$ can be derived, one corresponding to each intensity. However, the elucidation of the crystal structure requires also a knowledge of the complex numbers $E_H = |E_{\mathbf{H}}| \exp(i\phi_{\mathbf{H}})$, the normalized structure factors, of which only the magnitudes $|E_{\mathbf{H}}|$ can be determined from experiment. Thus a "phase" $\phi_{\mathbf{H}}$, unobtainable from the diffraction experiment, must be assigned to each $|E_{\mathbf{H}}|$, and the problem of determining the phases when only the magnitudes $|E_{\mathbf{H}}|$ are known is called the "phase problem". Owing to the known atomicity of crystal structures and the redundancy of observed magnitudes $|E_{\mathbf{H}}|$, the phase problem is solvable in principle.

2.2. *The structure invariants*

Equation (6) implies that the normalized structure factors E_H determine the crystal structure. However (5) does not imply that, conversely, the crystal structure determines the values of the normalized structure factors E_H since the position vectors $\mathbf{r}_j$ depend not only on the structure but on the choice of origin as well. It turns out nevertheless that the magnitudes $|E_{\mathbf{H}}|$ of the normalized structure factors are in fact uniquely determined by the crystal structure and are independent of the choice of origin but that the values of the phases $\phi_{\mathbf{H}}$ depend also on the choice of origin. Although the values of the individual phases depend on the structure and the choice of origin, there exist certain linear combinations of the phases, the so-called structure invariants, whose values are determined by the structure alone and are independent of the choice of origin.

It follows readily from Eq. (5) that the linear combination of three phases

$$\psi_3 = \phi_{\mathbf{H}} + \phi_{\mathbf{K}} + \phi_{\mathbf{L}} \tag{8}$$

is a structure invariant (triplet) provided that

$$\mathbf{H} + \mathbf{K} + \mathbf{L} = 0; \tag{9}$$

the linear combination of four phases

$$\psi_4 = \phi_{\mathbf{H}} + \phi_{\mathbf{K}} + \phi_{\mathbf{L}} + \phi_{\mathbf{M}} \tag{10}$$

is a structure invariant (quartet) provided that

$$\mathbf{H} + \mathbf{K} + \mathbf{L} + \mathbf{M} = 0\,; \tag{11}$$

etc.

2.3. The structure seminvariants

If a crystal possesses elements of symmetry then the origin may not be chosen arbitrarily if the simplifications permitted by the space group symmetries are to be realized. For example, if a crystal has a centre of symmetry it is natural to place the origin at such a centre while if a two-fold screw axis, but no other symmetry element is present, the origin would normally be situated on this symmetry axis. In such cases the permissible origins are greatly restricted and it is therefore plausible to assume that many linear combinations of the phases will remain unchanged in value when the origin is shifted only in the restricted ways allowed by the space group symmetries. One is thus led to the notion of the structure seminvariant, those linear combinations of the phases whose values are independent of the choice of permissible origin.

If the only symmetry element is a centre of symmetry, for example (space group $P\bar{1}$), then it turns out (again from Eq. (5)) that a single phase $\phi_{\mathbf{H}}$ is a structure seminvariant provided that the three components of the reciprocal lattice vector **H** are even integers; the linear combination of two phases $\phi_{\mathbf{H}}$ + $\phi_{\mathbf{K}}$ is a structure seminvariant provided that the three components of **H + K** are even integers; etc.

If the only symmetry element is a two-fold rotation axis (or twofold screw axis) then one finds from Eq. (5) that the single phase ϕ_{hkl} is a structure seminvariant provided that h and 1 are even integers and k = 0; the linear combination of two phases

$$\phi_{h_1k_1l_1} + \phi_{h_2k_2l_2}$$

is a structure seminvariant provided that $h_1 + h_2$ and $l_1 + l_2$ are even and $k_1 + k_2 = 0$; etc.

The structure invariants and seminvariants have been tabulated for all the space groups (Hauptman and Karle 1953, 1956, 1959; Karle and Hauptman 1961; Lessinger and Wondratschek 1975). In general the collection of structure invariants is a subset of the collection of structure seminvariants. If no element of symmetry is present, that is the space group is Pl, then the two classes coincide.

2.3.1. Origin and enantiomorph specification

The theory of the structure seminvariants leads in a natural way to space group dependent recipes for origin and enantiomorph (i.e. the handedness, right or left) specification.

In general the theory identifies an appropriate set of phases whose values are to be specified in order to fix the origin uniquely. For example, in space group Pl (no elements of symmetry) the values of any three phases

$$\phi_{h_1k_1l_1}, \phi_{h_2k_2l_2}, \phi_{h_3k_3l_3}, \quad (12)$$

for which the determinant Δ satisfies

$$\Delta = \begin{vmatrix} h_1k_1l_1 \\ h_2k_2l_2 \\ h_3k_3l_3 \end{vmatrix} = \pm 1, \quad (13)$$

may be specified arbitrarily, thus fixing the origin uniquely. Once this is done then the value of any other phase is uniquely determined by the structure alone. For enantiomorph specification it is sufficient to specify arbitrarily the sign of any enantiomorph sensitive structure invariant, i.e. one whose value is different from 0 or π. (See Hauptman 1972, pages 28-52, for further details.)

In the space group PI one again specifies arbitrarily the value (0 or π) of three phases (12), but now the condition is that the determinant Δ [defined by (13)] be odd. Similar recipes for all the space groups are now known and are to be found in the literature cited.

2.4. *The fundamental principle of direct methods*

It is known that the values of a sufficiently extensive set of cosine seminvariants (the cosines of the structure seminvariants) lead unambiguously to the values of the individual phases (Hauptman 1972). Magnitudes $|E|$ are capable of yielding estimates of the cosine seminvariants only or, equivalently, the magnitudes of the structure seminvariants; the signs of the structure seminvariants are ambiguous because the two enantiomorphous structures permitted by the observed magnitudes $|E|$ correspond to two values of each structure seminvariant differing only in sign. However, once the enantiomorph has been selected by specifying arbitrarily the sign of a particular enantiomorph sensitive structure seminvariant (i.e. one different from 0 or π), then the magnitudes $|E|$ determine both signs and magnitudes of the structure seminvariants consistent with the chosen enantiomorph. Thus, for fixed enantiomorph, the observed magnitudes $|E|$ determine unique values for the structure seminvariants; the latter, in turn, lead to unique values of the individual phases. In short, the structure seminvariants serve to link the observed magnitudes $|E|$ with the desired phases ϕ (the fundamental principle of direct methods). It is this property of the structure seminvariants which accounts for their importance and which justifies the stress placed on them here.

By the term "direct methods" is meant that class of methods which exploits relationships among the structure factors in order to go directly from the observed magnitudes $|E|$ to the needed phases ϕ.

2.5. The neighborhood principle

It has long been known that, for fixed enantiomorph, the value of any structure seminvariant ψ is, in general, uniquely determined by the magnitudes $|E|$ of the normalized structure factors. In recent years it has become clear that, for fixed enantiomorph, there corresponds to ψ one or more small sets of magnitudes $|E|$, the neighborhoods of ψ, on which, in favorable cases, the value of ψ most sensitively depends; that is to say that, in favorable cases, ψ is primarily determined by the values of $|E|$in any of its neighborhoods and is relatively insensitive to the values of the great bulk of remaining magnitudes. The conditional probability distribution of ψ, assuming as known the magnitudes $|E|$ in any of its neighborhoods, yields an estimate for ψ which is particularly good in the favorable case that the variance of the distribution happens to be small [the neighborhood principle (Hauptman, 1975a,b)].

The study of appropriate probability distributions (compare § 2.7) leads directly to the definition of the neighborhoods of the structure invariants. Definitions are given here only for the triplet ψ_3 and the quartet ψ_4, but recipes for defining the neighborhoods of all the structure invariants are now known (Hauptman 1977a,b, Fortier & Hauptman, 1977).

2.5.1. The first neighborhood of the triplet ψ_3

Let **H**, **K**, **L** be three reciprocal lattice vectors which satisfy Eq. (9). Then ψ_3, Eq. (8), is a structure invariant and its first neighborhood is defined to consist of the three magnitudes:

$$|E_{\mathbf{H}}|,\ |E_{\mathbf{K}}|,\ |E_{\mathbf{L}}|. \tag{14}$$

2.5.2. Neighborhoods of the quartet ψ_4

2.5.2.1. The first neighborhood

Let **H**, **K**, **L**, **M** be four reciprocal lattice vectors which satisfy Eq. (11). Then ψ_4, Eq. (10), is a structure invariant and its first neighborhood is defined to consist of the four magnitudes:

$$|E_{\mathbf{H}}|,\ |E_{\mathbf{K}}|,\ |E_{\mathbf{L}}|,\ |E_{\mathbf{M}}|. \tag{15}$$

The four magnitudes (15) are said to be the main terms of the quartet ψ_4.

2.5.2.2. The second neighborhood

The second neighborhood of the quartet ψ_4 is defined to consist of the four magnitudes (15) plus the three additional magnitudes:

$$|E_{\mathbf{H+K}}|,\ |E_{\mathbf{K+L}}|,\ |E_{\mathbf{L+H}}|, \tag{16}$$

i.e. seven magnitudes $|E|$ in all. The three magnitudes (16) are said to be the cross-terms of the quartet ψ_4.

2.6. The extension concept

By embedding the structure seminvariant **T** and its symmetry related variants in suitable structure invariants Q one obtains the extensions Q of the seminvariant T. Owing to the space group dependent relations among the phases, T is related in a known way to its extensions. In this way the theory of the structure seminvariants is reduced to that of the structure invariants. In particular, the neighborhoods of T are defined in terms of the neighborhoods of its extensions. The procedure will be illustrated in some detail only for the two-phase structure seminvariant in the space group PI which serves as the prototype for the structure seminvariants in general, in all space groups, noncentrosymmetric as well as centrosymmetric.

2.6.1. The two-phase structure seminvariant in $P\bar{1}$

It has already been seen (§ 2.3) that the linear combination of two phases

$$T = \phi_{\mathbf{H}} + \phi_{\mathbf{K}} \tag{17}$$

is a structure seminvariant in $P\bar{1}$ if and only if the three components of the reciprocal lattice vector $\mathbf{H} + \mathbf{K}$ are all even. Then the components of each of the four reciprocal lattice vectors $\frac{1}{2}(\pm \mathbf{H} \pm \mathbf{K})$ are all integers. Note also that in this space group the structure factors are real and all phases are O or π.

2.6.2. The extensions of T

One embeds the two-phase structure seminvariant T (17) and its symmetry related variant

$$T_1 = \phi_{-\mathbf{H}} + \phi_{\mathbf{K}} \tag{18}$$

in the respective quartets

$$Q = T + \phi_{-\frac{1}{2}(\mathbf{H}+\mathbf{K})} + \phi_{-\frac{1}{2}(\mathbf{H}+\mathbf{K})}, \tag{19}$$

$$Q_1 = T_1 + \phi_{-\frac{1}{2}(-\mathbf{H}+\mathbf{K})} + \phi_{-\frac{1}{2}(-\mathbf{H}+\mathbf{K})} \tag{20}$$

In view of (17) and (18) and the space group-dependent relationships among the phases it is readily verified that Q and Q_1 are in fact (special) four-phase structure invariants (quartets) and

$$T = T_1 = Q = Q_1. \tag{21}$$

The quartets Q and Q_1 are said to be the extensions of the seminvariant T. In this way the theory of the two-phase structure seminvariant T is reduced to that of the quartets. In particular, the neighborhoods of T are defined in terms of the neighborhoods of the quartet.

2.6.3. The first neighborhoods of the extensions

Since two of the phases of the quartet Q (19) are identical, only three of the four main terms are distinct. The first neighborhood of Q is accordingly defined to consist of the three magnitudes

$$\left[\text{since } \left|E_{-\frac{1}{2}(\mathbf{H}+\mathbf{K})}\right| = \left|E_{\frac{1}{2}(\mathbf{H}+\mathbf{K})}\right|\right]:$$

$$|E_{\mathbf{H}}|,\ |E_{\mathbf{K}}|,\ |E_{\frac{1}{2}(\mathbf{H}+\mathbf{K})}|. \tag{22}$$

In a similar way the first neighborhood of the extension Q_1, (20), is defined to consist of the three magnitudes

$$|E_{\mathbf{H}}|,\ |E_{\mathbf{K}}|,\ |E_{\frac{1}{2}(\mathbf{H}-\mathbf{K})}|. \tag{23}$$

2.6.4. The first neighborhood of T

The first neighborhood of the two-phase structure seminvariant T is defined to consist of the set-theoretic union of the first neighborhoods of its extensions, i.e., in view of (22) and (23), of the four magnitudes

$$|E_{\mathbf{H}}|,\ |E_{\mathbf{K}}|,\ |E_{\frac{1}{2}(\mathbf{H}+\mathbf{K})}|,\ |E_{\frac{1}{2}(\mathbf{H}-\mathbf{K})}|. \tag{24}$$

2.7. The solution strategy

One starts with the system of equations (5). By equating real and imaginary parts of (5) one obtains two equations for each reciprocal lattice vector H. The magnitudes $|E_H|$ and the atomic scattering factors f_j are presumed to be known. The unknowns are the atomic position vectors r_j and the phases $\phi_{\mathbf{H}}$. Owing to the redundancy of the system (5), one naturally invokes probabilistic techniques in order to eliminate the unknown position vectors r_j, and in this way to obtain relationships among the unknown phases $\phi_{\mathbf{H}}$ having probabilistic validity.

Choose a finite number of reciprocal lattice vectors **H**, **K**,... in such a way that the linear combination of phases

$$\psi = \phi_{\mathbf{H}} + \phi_{\mathbf{K}} + \ldots \tag{25}$$

is a structure invariant or seminvariant whose value we wish to estimate. Choose satellite reciprocal lattice vectors H′, K′, . . . in such a way that the collection of magnitudes

$$|E_{\mathbf{H}}|,\ |E_{\mathbf{K}}|,\ldots;\ |E_{\mathbf{H}'}|,\ |E_{\mathbf{K}'}|,\ldots \tag{26}$$

constitutes a neighborhood of ψ. The atomic position vectors r_j are assumed to be the primitive random variables which are uniformly and independently distributed. Then the magnitudes $|E_H|$, $|E_K|$,...; $|E_{H'}|$, $|E_{K'}|$,...; and phases $\phi_{\mathbf{H}}$, $\phi_{\mathbf{K}}$,...; $\phi_{\mathbf{H}'}$, $\phi_{\mathbf{K}'}$,... of the complex, normalized structure factors E_H, E_K,...; $E_{H'}$ $E_{K'}$,..., as functions [(Eq. (5)] of the position vectors r_j, are themselves random variables, and their joint probability distribution P may be obtained. From the distribution P one derives the conditional joint probability distribution

$$P\,(\Phi_{\mathbf{H}},\ \Phi_{\mathbf{K}},\ldots\ |\ |E_{\mathbf{H}}|,\ |E_{\mathbf{K}}|,\ldots;\ |E_{\mathbf{H'}}|,\ |E_{\mathbf{K'}}|,\ldots), \tag{27}$$

of the phases $\phi_{\mathbf{H}}, \phi_{\mathbf{K}},\ldots$, given the magnitudes $|E_H|$, $|E_K|,\ldots$; $|E_H'|$, $|E_K|,\ldots$, by fixing the known magnitudes, integrating with respect to the unknown phases $\phi_{\mathbf{H}}', \phi_{\mathbf{K}}',\ldots$ from O to 2π, and multiplying by a suitable normalizing parameter. The distribution (27) in turn leads directly to the conditional probability distribution

$$P(\Psi|\ |E_{\mathbf{H}}|,\ |E_{\mathbf{K}}|,\ldots;\ |E_{\mathbf{H'}}|,\ |E_{\mathbf{K'}}|,\ldots) \tag{28}$$

of the structure invariant or seminvariant ψ assuming as known the magnitudes (26) constituting a neighborhood of ψ. Finally, the distribution (28) yields an estimate for ψ which is particularly good in the favorable case that the variance of (28) happens to be small.

2.8. Estimating the triplet in PI

Let the three reciprocal lattice vectors **H**, **K**, and **L** satisfy (9). Refer to § 2.5.1 for the first neighborhood of the triplet ψ_3(Eq. (8)] and to § 2.7 for the probabilistic background.

Suppose that R_1, R_2, and R_3 are three specified non-negative numbers. Denote by

$$P_{1/3} = P(\Psi|R_1,\ R_2,\ R_3)$$

the conditional probability distribution of the triplet ψ_3, given the three magnitudes in its first neighborhood:

$$|E_{\mathbf{H}}| = R_1,\ |E_{\mathbf{K}}| = R_2,\ |E_{\mathbf{L}}| = R_3. \tag{29}$$

Then, carrying out the program described in § 2.7, one finds (Cochran, 1955)

$$P_{1/3} = P(\Psi|R_1,\ R_2,\ R_3) \approx \frac{1}{2\pi I_0(A)} \exp\,(A \cos \Psi) \tag{30}$$

where

$$A = \frac{2\sigma_3}{\sigma_2^{3/2}} R_1\, R_2\, R_3, \tag{31}$$

I_O is the modified Bessel function, and σ_n is defined by (7). Since $A>O$,$P_{1/3}$ has a unique maximum at $\psi = 0$, and it is clear that the larger the value of A the smaller is the variance of the distribution. See Figure 1, where A = 2.316, Figure 2, where A = 0.731. Hence in the favorable case that A is large, say, for example, A>3, the distribution leads to a reliable estimate of the structure invariant ψ_3, zero in this case:

$$\psi_3 \approx O \text{ if A is large.} \tag{32}$$

Furthermore, the larger the value of A, the more likely is the probabilistic statement (32). It is remarkable how useful this relationship has proven to be in the applications; and yet (32) is severely limited because it is capable of yielding only the zero estimate for ψ_3, and only those estimates are reliable for which A is large, the favorable cases.

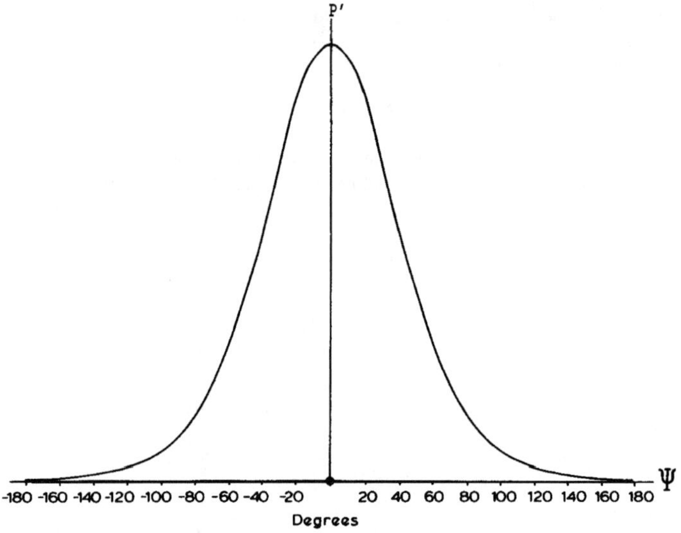

Figure 1. The distribution $P_{1/3}$, equation (30), for A = 2.316

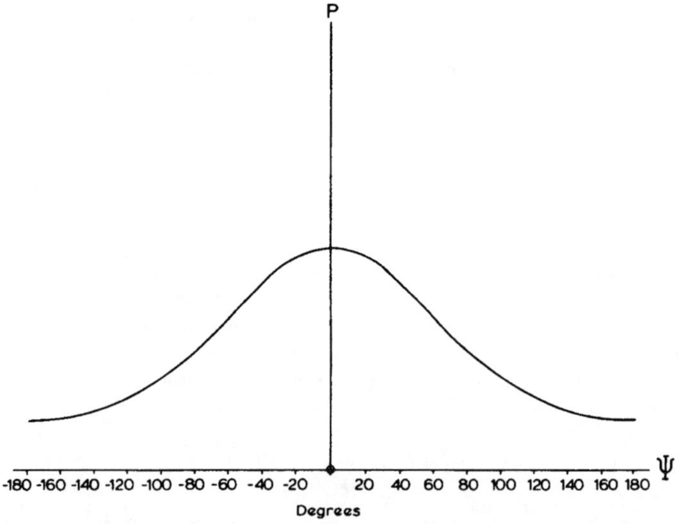

Figure 2. The distribution $P_{1/3}$, equation (30), for A = 0.731

It should be mentioned in passing that a distribution closely related to (30) leads directly to the so-called tangent formula (Karle & Hauptman, 1956) which is universaily used by direct methods practitioners:

$$\tan \phi_{\mathbf{h}} = \frac{\left\langle |E_{\mathbf{K}} E_{\mathbf{h}-\mathbf{K}}| \sin (\phi_{\mathbf{K}} + \phi_{\mathbf{h}-\mathbf{K}}) \right\rangle_{\mathbf{K}}}{\left\langle |E_{\mathbf{K}} E_{\mathbf{h}-\mathbf{K}}| \cos (\phi_{\mathbf{K}} + \phi_{\mathbf{h}-\mathbf{K}}) \right\rangle_{\mathbf{K}}} \tag{33}$$

in which **h** is a fixed reciprocal lattice vector, the averages are taken over the same set of vectors **K** in reciprocal space, usually restricted to those vectors **K** for which $|E_K|$ and $|E_{h-K}|$ are both large, and the sign of sin $\phi_{\mathbf{h}}$ (cos $\phi_{\mathbf{h}}$) is the same as the sign of the numerator (denominator) on the right hand side. The tangent formula is usually used to refine and extend a basis set of phases, presumed to be known.

2.9. Estimating the quartet in PI

Two conditional probability distributions are described, one assuming as known the four magnitudes $|E|$ in the first neighborhood of the quartet, the second assuming as known the seven magnitudes $|E|$ in its second neighborhood.

2.9.1. The first neighborhood

Suppose that **H**, **K**, **L**, and **M** are four reciprocal lattice vectors which satisfy (11). Refer to § 2.5.2.1 for the first neighborhood of the quartet ψ_4(10) and to § 2.7 for the probabilistic background. Suppose that R_1, R_2, R_3, and R_4 are four specified non-negative numbers. Denote by

$$P_{1/4} = P(\Psi | R_1, R_2, R_3, R_4)$$

the conditional probability distribution of the quartet ψ_4, given the four magnitudes in its first neighborhood:

$$|E_{\mathbf{H}}| = R_1, \ |E_{\mathbf{K}}| = R_2, \ |E_{\mathbf{L}}| = R_3, \ |E_{\mathbf{M}}| = R_4. \tag{34}$$

Then

$$P_{1/4} = P(\Psi | R_1, R_2, R_3, R_4) \approx \frac{1}{2\pi I_0(B)} \exp (B \cos \Psi) \tag{35}$$

where

$$B = \frac{2\sigma_4}{\sigma_2^2} R_1 R_2 R_3 R_4, \tag{36}$$

and $\sigma_{\mathbf{n}}$ is defined by (7). Thus $P_{1/4}$ is identical with P1/3, but B replaces A. Hence similar remarks apply to $P_{1/4}$. In particular, (35) always has a unique maximum at ψ = O so that the most probable value of ψ_4, given the four magnitudes (34) in its first neighborhood, is zero, and the larger the value of B the more likely it is that ψ_4 = O. Since B values, of order $1/N$, tend to be less than A values, of order $1/\sqrt{N}$, at least for large values of N, the estimate (zero) of ψ_4 is in general less reliable than the estimate (zero) of ψ_3. Hence the goal of obtaining a reliable non-zero estimate for a structure invariant is not realized by (35). The decisive step in this direction is made next.

2.9.2. The second neighborhood

Employ the same notation as in § 2.9.1 but refer now to § 2.5.2.2 for the second neighborhood of the quartet ψ_4. Suppose that R_1, R_2, R_3, R_4, R_{12}, R_{23}, and R_{31}, are seven non-negative numbers. Denote by

$$P_{1/7} = P(\Psi|R_1, R_2, R_3, R_4; R_{12}, R_{23}, R_{31})$$

the conditional probability distribution of the quartet ψ_4, given the seven magnitudes in its second neighborhood:

$$|E_{\mathbf{H}}| = R_1,\ |E_{\mathbf{K}}| = R_2,\ |E_{\mathbf{L}}| = R_3,\ |E_{\mathbf{M}}| = R_4; \tag{37}$$

$$|E_{\mathbf{H}+\mathbf{K}}| = R_{12},\ |E_{\mathbf{K}+\mathbf{L}}| = R_{23},\ |E_{\mathbf{L}+\mathbf{H}}| = R_{31}. \tag{38}$$

Then (Hauptman, 1975 a, b; 1976)

$$P_{1/7} \approx \frac{1}{L} \exp(-2B'\cos\Psi)\, I_0\left[\frac{2\sigma_3}{\sigma_2^{3/2}} R_{12}X_{12}\right] I_0\left[\frac{2\sigma_3}{\sigma_2^{3/2}} R_{23}X_{23}\right] \times$$

$$I_0\left[\frac{2\sigma_3}{\sigma_2^{3/2}} R_{31}X_{31}\right], \tag{39}$$

where

$$B' = \frac{1}{\sigma_2^3}(3\sigma_3^2 - \sigma_2\sigma_4)\, R_1R_2R_3R_4, \tag{40}$$

$$X_{12} = [R_1^2R_2^2 + R_3^2R_4^2 + 2R_1R_2R_3R_4\cos\Psi]^{1/2}, \tag{41}$$

$$X_{23} = [R_2^2R_3^2 + R_1^2R_4^2 + 2R_1R_2R_3R_4\cos\Psi]^{1/2}, \tag{42}$$

$$X_{31} = [R_3^2R_1^2 + R_2^2R_4^2 + 2R_1R_2R_3R_4\cos\Psi]^{1/2}, \tag{43}$$

σ_n is defined by (7), and L is a normalizing parameter, independent of ψ, which is not needed for the present purpose.

Figures 3-5 show the distribution (39) (solid line —) for typical values of the seven parameters (37) and (38). For comparison the distribution (35) (broken line ---) is also shown. Since the magnitudes |E| have been obtained from a real structure with N = 29, comparison with the true value of the quartet is also possible. As already emphasized, the distribution (35) always has a unique maximum at ψ = O. The distribution (39), on the other hand, may have a maximium at $\psi = O$, or π, or any value between these extremes, as shown by Figures 3-5. Roughly speaking, the maximum of (39) occurs at 0 or π according as the three parameters R_{12}, R_{23}, R_{31} are all large or all small, respectively. These figures also clearly show the improvement which may result when, in addition to the four magnitudes (37), the three magnitudes (38) are also assumed to be known. Finally, in the special case that

$$R_{12} \approx R_{23} \approx R_{31} \approx O \tag{44}$$

the distribution (39) reduces to

$$P_{1/7} \approx \frac{1}{L} \exp\left(-2B' \cos \Psi\right), \tag{45}$$

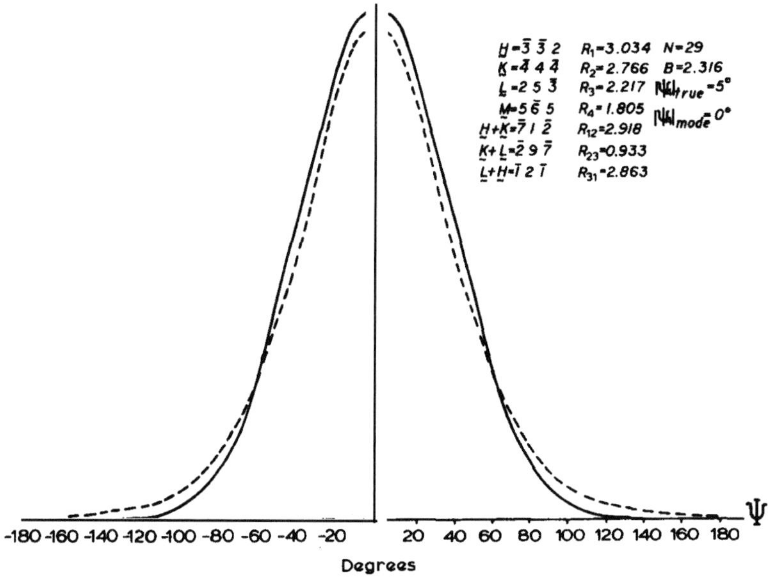

Figure 3. The distribution (39) (__) and (35) (---) for the values of the seven parameters (37) and (38) shown. The mode of (39) is O, of (35) always O.

APPENDIX E

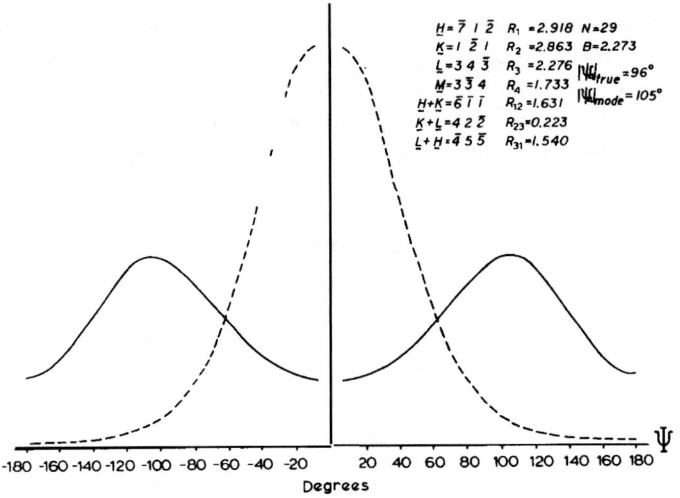

Figure 4. The distribution (39) (__) and (35) (---) for the values of the seven parameters (37) and (38) shown. The mode of (39) is 105°, of (35) always O.

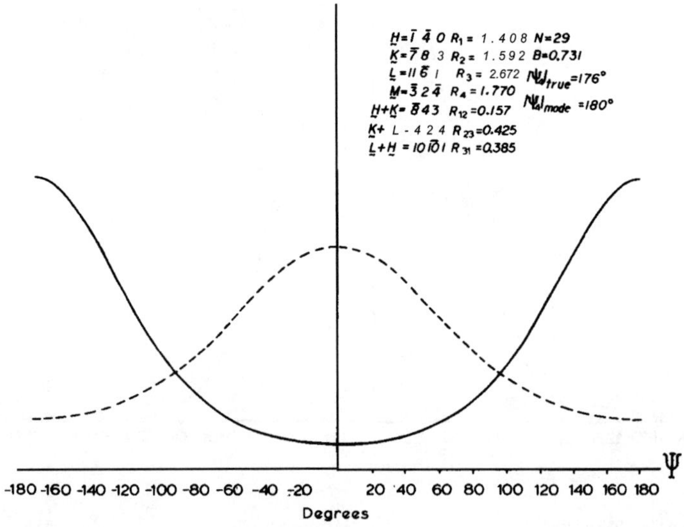

Figure 5. The distribution (39) (__) and (35) (---) for the values of the seven parameters (37) and (38) shown. The mode of (39) is 180°, of (35) always O.

which has a unique maximum at $\psi = \pi$ (Fig. 5).

2.10. Estimating the two-phase structure seminvariant in $P\bar{1}$

Suppose that **H** and **K** are two reciprocal lattice vectors such that the three components of **H** + **K** are even integers. Then the linear combination T of two phases (17) is a structure seminvariant. Refer to § 2.6.4 for the four magnitudes (24) in the first neighborhood of T and to § 2.7 for the probabilistic background. Suppose that R_1, R_2, r_{12}, and $r_{1\bar{2}}$ are four non-negative numbers. In this space group every phase is *O* or π so that T = 0 or π and the conditional probability distribution of T, assuming as known the four magnitudes in its first neighborhood, is discrete. Denote by $P_+(P_-)$ the conditional probability that T = *O* (π), given the four magnitudes in its first neighborhood:

$$|E_{\mathbf{H}}| = R_1,\ |E_{\mathbf{K}}| = R_2,\ |E_{\frac{1}{2}(\mathbf{H}+\mathbf{K})}| = r_{12},\ |E_{\frac{1}{2}(\mathbf{H}-\mathbf{K})}| = r_{1\bar{2}}. \tag{46}$$

In the special case that all N atoms in the unit cell are identical, the solution strategy described in § 2.7 leads to (Green & Hauptman, 1976)

$$P_{\pm} \approx \frac{1}{M} \exp\left\{ \mp \frac{R_1 R_2 \left(r_{12}^2 + r_{1\bar{2}}^2\right)}{2N} \right\} \cosh \left\{ \frac{r_{12} r_{1\bar{2}} \left(R_1 \pm R_2\right)}{N^{1/2}} \right\} \tag{47}$$

where upper (lower) signs go together and

$$M = \exp\left\{ -\frac{R_1 R_2 \left(r_{12}^2 + r_{1\bar{2}}^2\right)}{2N} \right\} \cosh \left\{ \frac{r_{12} r_{1\bar{2}} \left(R_1 + R_2\right)}{N^{1/2}} \right\} +$$

$$\exp\left\{ + \frac{R_1 R_2 \left(r_{12}^2 + r_{1\bar{2}}^2\right)}{2N} \right\} \cosh \left\{ \frac{r_{12} r_{1\bar{2}} \left(R_1 - R_2\right)}{N^{1/2}} \right\}. \tag{48}$$

It is easily verified that, under the assumption that R_1 and R_2 are both large, $P_+ >> 1/2$ if r_{12} and $r_{1\bar{2}}$ are both large, but $P_+ << 1/2$ if one of r_{12}, $r_{1\bar{2}}$ is large and the other is small. Hence T $\approx$ *O* or π respectively (the favorable cases of the neighborhood principle for T).

3. COMBINING DIRECT METHODS WITH ANOMALOUS DISPERSION

3.1. Introduction

The overview of the traditional direct methods described in §§ 1 and 2 is readily generalized to the case that the atomic scattering factors are arbitrary complex-valued functions of $(\sin \theta)/\lambda$, thus including the special case that one or more anomalous scatterers are present. Once again the neighborhood concept plays an essential role. Final results from the probabilistic theory of the two- and three-phase structure invariants are briefly summarized. In particular, the conditional probability distributions of the two- and three-phase structure invariants, given the magnitudes $|E|$ in their first neighborhoods, are described. The distributions yield estimates for these invariants which are particularly good in those cases that the variances of the distributions happen to be small (the neighborhood principle). It is particularly noteworthy that these estimates are unique in the whole range from $-\pi$ to $+\pi$. An example shows that the method is capable of yielding unique estimates for tens of thousands of structure invariants with unprecedented accuracy, even in the macromolecular case. It thus appears that this fusion of the traditional techniques of direct methods with anomalous dispersion will facilitate the solution of those crystal structures which contain one or more anomalous scatterers.

Most crystal structures containing as many as 80-100 independent nonhydrogen atoms are more or less routinely solvable nowadays by direct methods. On the other hand, it has been known for a long time (Peerdeman & Bijvoet, 1956; Ramachandran & Raman, 1956; Okaya & Pepinsky, 1956) that the presence of one or more anomalous scatterers facilitates the solution of the phase problem; and some recent work (Kroon, Spek & Krabbendam, 1977; Heinerman, Krabbendam, Kroon & Spek, 1978), employing Bijvoet inequalities and the double Patterson function, leads in a similar way to estimates of the sines of the three-phase structure invariants. Again, some early work of Rossmann (1961), employing the difference synthesis $(|F_H| - |F_{\bar{H}}|)^2$ in order to locate the anomalous scatterers and recently applied by Hendrickson and Teeter (1981) in their solution of the crambin structure, shows that the presence of anomalous scatterers facilitates the determination of crystal structures. This work strongly suggests that the ability to integrate the techniques of direct methods with anomalous dispersion would lead to improved methods for phase determination. The fusion of these techniques is described here. That the anticipated improvement is in fact realized is shown in Tables 1 and 2 and Fig. 6. Not only do the new formulas lead to improved estimates of the structure invariants but, more important still, because the distributions derived here are unimodal in the whole interval from $-\pi$ to $+\pi$, the twofold ambiguity inherent in all the earlier work is removed. It is believed that this resolution of the twofold ambiguity results from the ability now to make use of the individual magnitudes in the first neighborhood of the structure invariant and the avoidance of explicit dependence on the Bijvoet differences; the explicit use of the Bijvoet differences, as had been done in all previous work, leads apparently to a loss of information resulting in a twofold ambiguity in estimates of the structure invariants. It may be of some interest to observe that in the earlier work with anomalous dispersion only the sine of the invariant may be estimated; in the absence of anomalous scatterers only the cosine of the invariant may be estimated; as a result of the work described here both the sine and the cosine, that is to say the invariant itself, may be estimated. Since, in the presence of anomalous scatterers, the observed intensities are known to determine a unique enantiomorph, and therefore unique values for all the structure seminvariants, formulas of the kind described here should not be unexpected; nevertheless not even their existence appears to have been anticipated.

3.2. The normalized structure factors

In the presence of anomalous scatterers the normalized structure factor

$$E_{\mathbf{H}} = |E_{\mathbf{H}}|\exp(i\phi_{\mathbf{H}}) \tag{49}$$

is defined by

$$E_{\mathbf{H}} = \frac{1}{\alpha_{\mathbf{H}}^{1/2}} \sum_{j=1}^{N} f_{j\mathbf{H}} \exp(2\pi i \mathbf{H}\cdot\mathbf{r}_j) \tag{50}$$

$$= \frac{1}{\alpha_{\mathbf{H}}^{1/2}} \sum_{j=1}^{N} |f_{j\mathbf{H}}| \exp[i(\delta_{j\mathbf{H}} + 2\pi \mathbf{H}\cdot\mathbf{r}_j)] \tag{51}$$

where

$$f_{j\mathbf{H}} = |f_{j\mathbf{H}}|\exp(i\delta_{j\mathbf{H}}) \tag{52}$$

is the (in general complex) atomic scattering factor (a function of $|\mathbf{H}|$ as well as of j) of the atom labeled j, $\mathbf{r}_j$ is its position vector, N is the number of atoms in the unit cell, and

$$\alpha_{\mathbf{H}} = \sum_{j=1}^{N} |f_{j\mathbf{H}}|^2. \tag{53}$$

For a normal scatter, $\delta_{j\mathbf{H}} = O$; for an atom which scatters anomalously, $\delta_{j\mathbf{H}} \neq O$. Owing to the presence of the anomalous scatterers, the atomic scattering factors $f_{j\mathbf{H}}$, as functions of sin θ/λ, do not have the same shape for different atoms, even approximately. Hence the dependence of the f_{jH} on $|\mathbf{H}|$ cannot be ignored, in contrast to the usual practice when anomalous scatterers are not present. For this reason the subscript **H** is not suppressed in the symbols fj_{H} and $\alpha_{\mathbf{H}}$ [Eq. (53)].

The reciprocal-lattice vector **H** is assumed to be fixed, and the primitive random variables are taken to be the atomic position vectors r_j which are assumed to be uniformly and independently distributed. Then E_{H}, as a function, (Eq. (51)], of the primitive random variables r_j, is itself a random variable and, as it turns out,

$$\left\langle |E_{\mathbf{H}}|^2 \right\rangle_{\mathbf{r}_j} = 1. \tag{54}$$

3.3. The two-phase structure invariant

The two-phase structure invariant, which has no analogue when no anomalous scatterers are present, is defined by

$$\psi = \phi_{\mathbf{H}} + \phi_{\bar{\mathbf{H}}}. \tag{55}$$

APPENDIX E

3.3.1. The first neighborhood

The first neighborhood of the two-phase structure invariant ψ [Eq. (55)] is defined to consist of the two magnitudes

$$|E_{\mathbf{H}}|, |E_{\bar{\mathbf{H}}}|, \tag{55}$$

which, because of the breakdown of Friedel's Law, are in general distinct.

3.3.2. Estimating the two-phase structure invariant

Define C_H and S_H by means of

$$C_{\mathbf{H}} = \frac{1}{\alpha_{\mathbf{H}}} \sum_{j=1}^{N} |f_{j\mathbf{H}}|^2 \cos 2\,\delta_{j\mathbf{H}} \tag{57}$$

$$S_{\mathbf{H}} = \frac{1}{\alpha_{\mathbf{H}}} \sum_{j=1}^{N} |f_{j\mathbf{H}}|^2 \sin 2\,\delta_{j\mathbf{H}} \tag{58}$$

where f_{jH}, $\delta_{j\mathbf{H}}$, and $\alpha_{\mathbf{H}}$ are defined in (52) and (53). Define X and ξ by means of

$$X \cos \xi = C_{\mathbf{H}},\ X \sin \xi = -S_{\mathbf{H}}, \tag{59}$$

$$X = \left[C_{\mathbf{H}}^2 + S_{\mathbf{H}}^2\right]^{1/2},\ \tan \xi = -S_{\mathbf{H}}/C_{\mathbf{H}} \tag{60}$$

Suppose that R and $\bar{R}$ are fixed non-negative numbers. In view of (48) to (50) the two-phase structure invariant $\phi_{\mathbf{H}} + \phi_{\bar{\mathbf{H}}}$, as a function of the primitive random variables r_j, is itself a random variable. Denote by $P(\Psi|R,\bar{R})$ the conditional probability distribution of the two-phase structure invariant $\phi_{\mathbf{H}} + \phi_{\bar{\mathbf{H}}}$, given the two magnitudes in its first neighborhood:

$$|E_{\mathbf{H}}| = R,\ |E_{\bar{\mathbf{H}}}| = \bar{R}. \tag{61}$$

Then (Hauptman, 1982; Giacovazzo, 1983)

$$P(\Psi|R,\bar{R}) \approx \left[2\pi I_0\left[\frac{2R\bar{R}X}{1-X^2}\right]\right]^{-1} \exp\left\{\frac{2R\bar{R}X}{1-X^2}\cos(\Psi+\xi)\right\}, \tag{62}$$

where X and ξ, defined by (57) - (60) are seen to be functions of the (complex) atomic scattering factors $f_{j\mathbf{H}}$, which are presumed to be known. It should be noted that the distribution (62) has the same form as (30) but is centered at $-\xi$ instead of *O*. Since (62) has a unique maximum at $\Psi = -\xi$, it follows that

$$\phi_H + \phi_{\bar{H}} \approx -\xi \tag{63}$$

provided that the variance of the distribution is small i.e. provided that

$$A = \frac{2R\bar{R}X}{1-X^2} \text{ is large.} \tag{64}$$

It should be noted that, while A depends on R, $\bar{R}$ and $|\mathbf{H}|$, for a fixed chemical composition ξ depends only on $|\mathbf{H}|$ (or $\sin\theta/\lambda$) and is independent of R and $\bar{R}$.

3.4. The three-phase structure invariant

It will be assumed throughout that **H, K,** and **L** are fixed reciprocal-lattice vectors satisfying

$$\mathbf{H} + \mathbf{K} + \mathbf{L} = \mathbf{O}\,. \tag{65}$$

Owing to the breakdown of Friedel's law there are, in sharp contrast to the case that no anomalous scatterers are present, eight distinct three-phase structure invariants:

$$\psi_0 = \phi_{\mathbf{H}} + \phi_{\mathbf{K}} + \phi_{\mathbf{L}}, \tag{66}$$

$$\psi_1 = -\,\phi_{\bar{\mathbf{H}}} + \phi_{\mathbf{K}} + \phi_{\mathbf{L}}, \tag{67}$$

$$\psi_2 = \phi_{\mathbf{H}} - \phi_{\bar{\mathbf{K}}} + \phi_{\mathbf{L}}, \tag{68}$$

$$\psi_3 = \phi_{\mathbf{H}} + \phi_{\mathbf{K}} - \phi_{\bar{\mathbf{L}}}, \tag{69}$$

$$\psi_{\bar{0}} = \phi_{\bar{\mathbf{H}}} + \phi_{\bar{\mathbf{K}}} + \phi_{\bar{\mathbf{L}}}, \tag{70}$$

$$\psi_{\bar{1}} = -\phi_{\mathbf{H}} + \phi_{\bar{\mathbf{K}}} + \phi_{\bar{\mathbf{L}}}, \tag{71}$$

$$\psi_{\bar{2}} = \phi_{\bar{\mathbf{H}}} - \phi_{\mathbf{K}} + \phi_{\bar{\mathbf{L}}}, \tag{72}$$

$$\psi_{\bar{3}} = \phi_{\bar{\mathbf{H}}} + \phi_{\bar{\mathbf{K}}} - \phi_{\mathbf{L}}. \tag{73}$$

3.4.1. The first neighborhood

The first neighborhood of each of the three-phase structure invariants (66)-(73) is defined to consist of the six magnitudes:

$$|E_{\mathbf{H}}|,\ |E_{\mathbf{K}}|,\ |E_{\mathbf{L}}|,\ |E_{\bar{\mathbf{H}}}|,\ |E_{\bar{\mathbf{K}}}|,\ |E_{\bar{\mathbf{L}}}| \tag{74}$$

which, again owing to the breakdown of Friedel's law, are not in general equal in pairs.

3.4.2. The probabilistic background

Fix the reciprocal-lattice vectors **H**, **K**, and **L**, subject to (65). Suppose that the six non-negative numbers R_1, R_2, R_3, $R_{\bar{1}}$, $R_{\bar{2}}$ and $R_{\bar{3}}$ are also specified. Define the N-fold Cartesian product W to consist of all ordered N-tuples $(r_1, r_2, \ldots, r_N)$, where $r_1, r_2, \ldots, r_N$ are atomic position vectors. Suppose that the primitive random variable is the N-tuple $(r_1, r_2, \ldots, r_N)$ which is assumed to be uniformly distributed over the subset of W defined by

$$|E_{\mathbf{H}}| = R_1,\ |E_{\mathbf{K}}| = R_2,\ |E_{\mathbf{L}}| = R_3, \tag{75}$$

$$|E_{\bar{\mathbf{H}}}| = R_{\bar{1}},\ |E_{\bar{\mathbf{K}}}| = R_{\bar{2}},\ |E_{\bar{\mathbf{L}}}| = R_{\bar{3}}, \tag{76}$$

where the normalized structure factors E are defined by (50). Then the eight structure invariants

$$\psi_j,\ \psi_{\bar{j}},\ j = 0, 1, 2, 3, \tag{77}$$

(66)-(73), as functions of the primitive random variables $(r_1, r_2, \ldots r_N)$, are themselves random variables.

Our major goal is to determine the conditional probability distribution of each of the three-phase structure invariants (66)-(73), given the six magnitudes (75) and (76) in its first neighborhood, which, in the favorable case that the variance of the distribution happens to be small, yields a reliable estimate of the invariant (the neighborhood principle).

3.4.3. Estimating the three-phase structure invariant

Denote by

$$P_j(\Psi|R_1, R_2, R_3, R_{\bar{1}}, R_{\bar{2}}, R_{\bar{3}}) = P_j(\Psi),$$

$$j = 0, 1, 2, 3, \bar{0}, \bar{1}, \bar{2}, \bar{3}, \tag{78}$$

the conditional probability distribution of each ψ_j, assuming as known the six magnitudes (74) in its first neighborhood. Then the final formula, the major result of this article, is simply (Hauptman, 1982; Giacovazzo, 1983)

$$P_j(\Psi) \approx \frac{1}{K_j} \exp\left\{ A_j \cos(\Psi - \omega_j) \right\},$$

$$j = 0, 1, 2, 3, \bar{0}, \bar{1}, \bar{2}, \bar{3} \tag{79}$$

where the parameters K_j, A_j, and ω_j are expressible in terms of the complex scattering factors f_{jH}, f_{jK}, f_{jL}, presumed to be known, and the observed magnitudes $|E_H|$, $|E_K|$, $|E_L|$, $|E_{\bar{H}}|$, $|E_{\bar{K}}|$, $|E_{\bar{L}}|$ in the first neighborhood of the invariant. Since the K_j's and A_j's are positive, the maximum of (79) occurs at $\Psi = \omega_j$. Hence when the variance of the distribution (79) is small, i.e. when A_j is large, one obtains the reliable estimate

$$\psi_j = \omega_j, j = 0, 1, 2, 3, \bar{0}, \bar{1}, \bar{2}, \bar{3}, \tag{80}$$

for the structure invariant ψ_j. It should be emphasized that the estimate (80) is unique in the whole range from $-\pi$ to $+\pi$. No prior knowledge of the positions of the anomalous scatterers is needed, nor is it required that the anomalous scatterers be identical.

3.4.4. The applications

Using the presumed known coordinates of the $PtCl_4^{2-}$ derivative of the protein Cytochrome c_{550} from *Paracoccus denitrificans* (Timkovich & Dickerson, 1976), molecular weight $M_r \simeq$ 14,500, space group $P2_12_12_1$, some 8300 normalized structure factors E were calculated (to a resolution of 2.5Å). In addition to the anomalous scatterers Pt and Cl, this structure contains one Fe and six S atoms which also scatter anomalously at the wavelength used (CuKα). Using the 4000 phases ϕ_{hkl} corresponding to the 4000 largest $|E_{hkl}|$'s with hkl ≠ O, the three-phase structure invariants ψ_j, j = 0, 1, 2, 3, $\bar{0}$, $\bar{1}$, $\bar{2}$, $\bar{3}$, [(66)-(73)], were generated and the parameters ω_j and A_j needed to define the distributions (79), were calculated. All calculations were done on the VAX 11/780 computer; double precision (approximately 15 significant digits) was used in order to eliminate round-off errors. The values of the A_j's were arranged in descending order and the first 2000, sampled at intervals of 100, were used in the construction of Table 1; the top 60,000 were used for Table 2.

Table 1 lists 21 values of A_j, sampled as shown from the top 2000, the corresponding estimates ω_j (in degrees) of the invariants ψ_j, the true values of the ψ_j, and the magnitude of the error, $|\omega_j-\psi_j|$. Also listed are the six magnitudes $|E|$ in the first neighborhood of the corresponding invariant.

Table 2 gives the average magnitude of the error,

$$<|\omega_j - \psi_j|>, \tag{81}$$

in the nine cumulative groups shown, for the 60,000 most reliable estimates ω_j of the invariants ψ_j.

Tables 1 and 2 show firstly that, owing to the unexpectedly large number of large values of A_j, our formulas yield reliable (and unique) estimates of tens of thousands of the three-phase structure invariants. Secondly, the invariants which are most reliably estimated lie anywhere in the range from -180" to +180°, and appear to be uniformly distributed in this range (Columns 9 and 10 of Table 1). Finally , in sharp contrast to the case that no anomalous scatterers are present, the most reliable estimates are not necessarily of invariants corresponding to the most intense reflections but of those corresponding instead to reflections of only moderate intensity (Columns 2-7 of Table 1).

APPENDIX E

Fig. 6 shows a scatter diagram of ω_j versus ψ_j for the $PtCl_4^{2-}$ derivative of Cytochrome c_{550}, using 201 invariants sampled at intervals of length ten from the top 2000, as well as the line $\omega_j = \psi_j$. Since the line falls evenly among the points, it appears that the ω_j are unbiased estimates of the invariants ψ_j.

3.4.5. Concluding remarks

In this article the goal of integrating the techniques of direct methods with anomalous dispersion is realized. Specifically, the conditional probability distribution of the three-phase structure invariant, assuming as known the six magnitudes in its first neighborhood, is obtained. In the favorable case that the variance of the distribution happens to be small, the distribution yields a reliable estimate of the invariant (the neighborhood principle). It is particularly noteworthy that, in strong contrast to all previous work, the estimate is unique in the whole interval $(-\pi, \pi)$ and that any estimate in this range is possible (even, for example, in thr vicinity of $\pm\pi/2$ or π). The first applications of this work using error-free diffraction data have been made, and these show that in a typical case some tens of thousands of three-phase structure invariants may be estimated with unprecedented accuracy, even for a macromolecular crystal structure. Some preliminary calculations on a number of structures, not detailed here, show that the accuracy of the estimates depends in some complicated way on the complexity of the crystal structure, the number of anomalous scatterers, the strength of the anomalous signal, and the range of sin θ/λ. With smaller structures the accuracy may be greatly increased, average errors of only three or four degrees for thousands of invariants being not uncommon.

It should be stated in conclusion that the availability of reliable estimates for large numbers of the three-phase structure invariants implies that the traditional machinery of direct methods, in particular the tangent formula [Eq. (33)], suitably modified to accommodate the non-zero estimates of the invariants, may be carried over without essential change to estimate the values of the individual phases and thus to facilitate structure determination via anomalous dispersion. In view of the calculations summarized in Tables 1 and 2 and Fig. 6, it seems likely that, in time, even macromolecules will prove to be solvable in this way. It is clear, too, that, owing to the ability to estimate both the sine and cosine invariants, that is to say both the signs and magnitudes of the invariants, the unique enantiomorph determined by the observed intensities is automatically obtained. In fact the first application of these techniques using experimental diffraction data has already facilitated the solution of the unknown macromolecular structure Cd, Zn Metallothionein [Furey, et al., 1986].

APPENDIX E

Table 1. Twenty-one estimates ω_j (in degrees) of the structure invariants ψ_j sampled from the top 2,000 for the Pt Cl_4^{2-} derivative of Cytochrome c_{550}.

Serial No.	$\|E_H\|$	$\|E_{\bar{H}}\|$	$\|E_K\|$	$\|E_{\bar{K}}\|$	$\|E_L\|$	$\|E_{\bar{L}}\|$	A_j	Estimated value ω_j of ψ_j	True value of ψ_j	Mag. of the Error $\|\omega_j-\psi_j\|$
1	2.17	2.04	0.89	1.03	0.85	0.67	6.92	− 58°	− 88°	30°
100	1.91	2.06	1.61	1.49	0.85	0.67	5.62	148	130	18
200	1.91	2.06	1.96	2.06	1.41	1.57	4.83	− 79	−121	42
300	2.36	2.48	1.56	1.69	0.82	0.68	4.52	52	2	50
400	2.17	2.04	1.34	1.48	1.28	1.15	4.31	79	96	17
500	1.85	1.94	0.85	0.67	0.78	0.92	4.21	56	42	14
600	2.17	2.04	0.92	1.04	0.86	0.70	4.10	146	148	2
700	1.39	1.28	0.85	0.67	0.87	0.75	4.02	− 72	− 68	4
800	1.41	1.57	1.61	1.49	0.71	0.85	3.93	70	50	20
900	1.88	1.98	1.28	1.15	0.85	0.67	3.87	104	96	8
1,000	1.29	1.43	0.79	0.71	0.85	0.67	3.80	− 88	−138	50
1,100	1.34	1.48	1.34	1.22	1.25	1.16	3.76	− 72	−126	54
1,200	1.56	1.69	1.41	1.57	0.98	0.90	3.72	73	78	5
1,300	1.98	2.07	2.08	1.94	1.08	1.21	3.68	−161	−124	37
1,400	1.56	1.67	1.41	1.57	1.24	1.33	3.63	− 72	− 3	69
1,500	2.38	2.50	1.91	2.06	0.74	0.64	3.59	84	77	7
1,600	1.91	2.06	1.34	1.22	0.72	0.83	3.55	− 64	− 94	30
1,700	1.91	2.06	2.02	2.12	2.15	2.24	3.51	− 64	− 72	8
1,800	2.38	2.50	1.61	1.49	0.78	0.90	3.46	78	82	4
1,900	2.38	2.50	1.63	1.70	1.81	1.93	3.43	63	123	60
2,000	0.85	0.67	0.97	0.83	1.02	1.09	3.42	− 96	−126	30

APPENDIX E

Table 2. Average magnitude of the error (in degrees) in the top 60,000 estimated values of the three-phase structure invariants, cumulated in the nine groups shown, for the Pt Cl_4^{2-} derivative of Cytochrome c_{550}.

Group No.	No. in Group	Average Value of A	Average Mag. of Error
1	100	6.01	27.9°
2	500	4.90	29.3
3	1,000	4.44	28.8
4	2,000	4.02	28.0
5	5,000	3.49	31.4
6	10,000	3.09	33.8
7	20,000	2.71	36.1
8	40,000	2.35	38.6
9	60,000	2.15	39.8

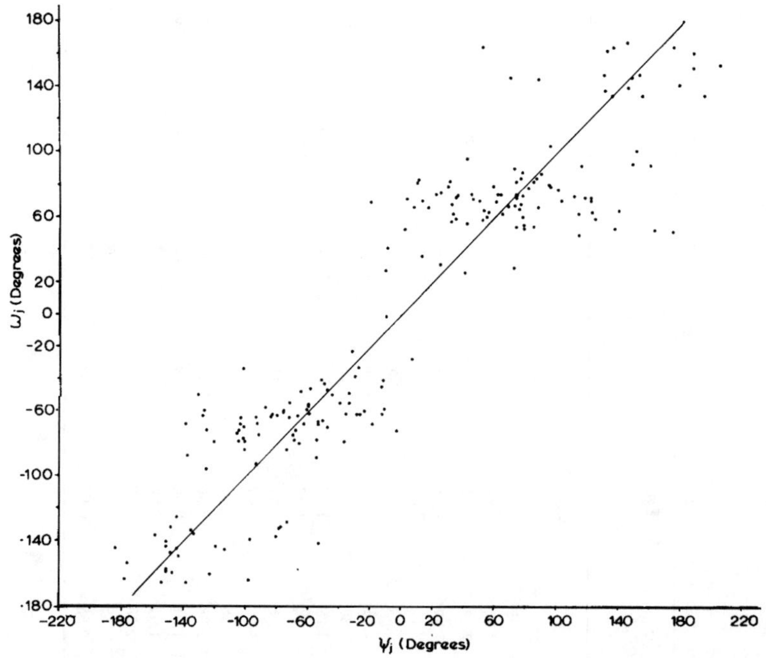

Fig. 6. A scatter diagram of ω_j *versus* ψ_j, using 201 invariants sampled at intervals of length ten from the top 2000, for the $PtCl_4^{2-}$ derivative of Cytochrome c_{550}, as well as the line $\omega_j = \psi_j$.

APPENDIX E

REFERENCES

Cochran, W. (1955), Acta Cryst. 8, 473-478.

Green, E. and Hauptman, H. (1976), Acta Cryst. A32, 940-944.

Fortier, S. and Hauptman, H. (1977), A33, 694-696.

Furey, W. F., Robbins, A. H., Clancy, L. L., Winge, D. R., Wang, B. C., and Stout, C. D. (1986). Science 231, 704-710.

Giacovazzo, C. (1983), Acta Cryst. A39, 585-592.

Hauptman, H. (1972), Crystal Structure Determination: The Role of the Cosine Seminvariants (New York: Plenum Press).

- (1975a), Acta Cryst. A31, 671-679.

- (1975b), Acta Cryst. A31, 680-687.

- (1976), Acta Cryst. A32, 877-882.

- (1977a), Acta Cryst. A33, 553-555.

- (1977b), Acta Cryst. A33 , 568-571.

- (1982), Acta Cryst. A38 , 632-641.

- and Karle, J. (1953), Solution of the Phase Problem I. The Centrosymmetric Crystal. ACA Monograph No. 3. Polycrystal Book Service.

- and Karle, J. (1956), Acta Cryst. 9, 45-55.

- and Karle, J. (1959), Acta Cryst. 12, 93-97.

Heinerman, J. J. L., Krabbendam, H., Kroon, J., and Spek, A. L. (1978). Acta Cryst. A34, 447-450.

Hendrickson, W. A. and Teeter, M. M. (1981), Nature (London) 290, 107- 113.

Karle, J. and Hauptman, H. (1955), Acta Cryst. 9, 635-651.

- - (1961), Acta Cryst. 14, 217-223.

Kroon, J., Spek, A. L., and Krabbendam, H. (1977), Acta Cryst. A33,382-385.

Lessinger, L. and Wondratschek, H. (1975), Acta Cryst. A31, 521.

Okaya, Y. and Pepinsky, R. (1956), Phys. Rev. 103, 1645-1647.

Peerdeman, A. F. and Bijvoet, J. M. (1956), Proc. K. Ned. Akad. Wet. B59, 312- 313.

Ramachandran, G. N. and Raman, S. (1956), Curr. Sci. 25, 348-351.

Rossmann, M. G. (1961), Acta Cryst. 14, 383-388 .

Timkovich, R. and Dickerson, R. E. (1976), J. Biol. Chem. 251, 4033-4046 .

Herbert Hauptman during World War II.
(Courtesy of Hauptman-Woodward Institute)

Presented with the Nobel Prize in Chemistry in 1985, Stockholm, Sweden. *(Courtesy of Hauptman-Woodward Institute)*

Showing a crystal structure.
(Courtesy of Hauptman-Woodward Institute)

Another crystal structure.
(Courtesy of Hauptman-Woodward Institute)

Herbert Hauptman in his office at the Hauptman-Woodward Institute, Buffalo, New York. *(Courtesy of Hauptman-Woodward Institute)*

With Buffalo mayor Byron Brown, 2005.
(Courtesy of Hauptman-Woodward Institute)

Appendix F

ORIGINAL PAPER DETAILING THE DISCOVERY THAT WON THE NOBEL PRIZE

Solution of the Phase Problem

ACA MONOGRAPH
Number 3

SOLUTION OF THE PHASE PROBLEM
I. THE CENTROSYMMETRIC CRYSTAL

By
Herbert Hauptman
and
Jerome Karle

Naval Research Laboratory

Published by AMERICAN CRYSTALLOGRAPHIC ASSOCIATION
Successor to AMERICAN SOCIETY FOR X-RAY AND ELECTRON DIFFFRACTION (ASXRED) and AMERICAN CRYSTALLOGRAPHIC SOCIETY (ACS)

September, 1953

Errata for ACA Monograph Number 3, Solution of the Phase Problem

I. The Centrosymmetric Crystal

H. Hauptman and J. Karle

P. 7. (1.27) should read: $E = F/(\frac{m2}{n}\sigma_2)^{1/2}$

P. 13. line 18. $y_j = \mathcal{E}_2$ should read: $y_j - \mathcal{E}_2$

P. 51. Last sentence in heading of Table 10 should read: Case 2 does not include Case 1.

Pp.52,53. Last sentences in headings of Tables 11 and 12 should read: No case includes a previous case.

P. 53. Table 12. Last sentence on last line should read: If $k_\mu = \pm k_\nu$, the entries should read $+\frac{3}{8}\sqrt{2}$, $-\frac{1}{8}\sqrt{2}$, $-\frac{1}{8}\sqrt{2}$, $+\frac{3}{8}\sqrt{2}$ in order.

P. 63. In denominator of (4.40) replace exponent 3/2 by 4/2.

P. 66. In denominator of coefficient of $\sum'_{ji}$ of (4.43) replace exponent 3/2 by 4/2.

P. 63. Eq. (4.41) might be clearer if $\sum_{\nu}$ were replaced by $\sum_{k_\mu}$.

P. 66. Line following (4.44) should read $\vec{h}_\mu = \pm\frac{1}{2}\vec{h}_i$.

P. 68. Eq. (4.48). $(-1)^{h_\mu + k_\mu + \varepsilon}$ should read $(-1)^{\varepsilon}$.

P. 68. Eq. (4.49). Replace $h_i \pm h_j$ by $\frac{1}{2}(h_i \pm h_j)$.

P. 68. Eq. (4.50). Replace $l_i \pm l_j$ by $\frac{1}{2}(l_i \pm l_j)$.

P. 68. Eq. (4.51). Replace $k_i \pm k_j$ by $\frac{1}{2}(k_i \pm k_j)$.

P. 70. Eq. (4.53). Replace $h_i \pm h_j$ by $\frac{1}{2}(h_i \pm h_j)$, and replace $l_i \pm l_j$ by $\frac{1}{2}(l_i \pm l_j)$.

P. 71. Legend for Fig. 2. Replace h by h_μ, k by k_μ, l by l_μ.

P. 72. Legend for Fig. 3. Replace h by h_μ, k by k_μ, l by l_μ.

P. 83. Line 7. Replace kx + ky + by hx + ky + .

P. 52. Line 4 of Table 11. $h_\mu = l_\nu = 0$ should read: $h_\mu = l_\mu = 0$.

To Our Parents

PREFACE

This Monograph deals with the application of probability methods to the study of a very complicated system of simultaneous equations, the structure factor equations for centrosymmetric crystals. In this set of equations the atomic structure factors and the magnitudes of the crystal structure factors are assumed to be known. The phases (or signs) of the crystal structure factors and the atomic coordinates are regarded as unknown. The main purpose of this Monograph is the derivation, in Chapter 4, of a routine procedure for determining the phases of the structure factors which is valid for all centrosymmetric space groups. Since the system of structure factor equations is in general redundant, (i.e. there are more equations than are needed to determine the solution in a strict algebraic sense) and since the magnitudes of the structure factors, determined from experiment, are subject to experimental errors, it is natural to invoke probability methods in the search for a solution of the problem. For these reasons our solution has the character of a least squares solution. It is a "most probable" solution, or a solution which "best" fits the redundant experimental data. Although the statistical methods used necessarily attach a probability measure to the final results, ordinarily the problem is so overdetermined by the amount of experimental data available, that these probability measures are often close to unity.

It is well known that the phase of a structure factor depends not only on the crystal structure but also on the choice of origin (see e.g. the recent work of Okaya and Nitta, 1952). The exact nature of this dependence is clarified in Chapter 2 for all centrosymmetric space groups by means of the introduction of special linear combinations of the phases, the structure invariants and seminvariants. A crucial test of the probability theory is whether it leads to the same dependence of phase on the choice of origin as that derived from the analysis of the invariants and seminvariants. How this comes about in a decidedly non-trivial fashion is shown in some detail for space groups $P\bar{1}$ and $P2_1/a$, and is briefly indicated for space groups $P4/m$, $P\bar{3}$, and $R\bar{3}$. This agreement is considered to be important evidence in support of the validity of the probability theory.

The basic concepts and methods underlying the probability approach were developed in three previous papers. In the first of these (Hauptman and Karle, 1952) the structure factor equations were interpreted as coupled, plane, closed, vector polygons. With this interpretation the solution to the problem of the random walk was used to determine the probability distribution of the magnitude of a structure factor. From this the probability distributions of the interatomic vectors could be inferred. It was found necessary, however, to treat the coupling among the vector polygons in an approximate manner. Furthermore, the methods used became increasingly unwieldy when attempts were made to make use of the additional symmetry of the more complicated space groups. Finally, since only the interatomic vectors were obtained, there still remained the problem of finding the atomic coordinates. The shortcomings of this geometric approach made it desirable to develop analytic tools capable of treating the coupling among the vector polygons in an exact manner no matter how complicated the crystal symmetry. These tools have been developed in the two succeeding papers, which include significant results related to the original problem, and in the present Monograph. While the geometric interpretation is no longer needed to obtain useful results, it nevertheless remains in the background and motivates the analytical development.

In the second paper (Karle and Hauptman, 1953a) probability distributions were obtained for a structure factor (and its magnitude) for every centrosymmetric space group. These were obtained by fixing a vector $\vec{h}$ and allowing the atoms in the asymmetric unit to range uniformly and independently throughout the unit cell. The fact was emphasized that these distributions are conceptually distinct from, but in general identical with, those obtained by fixing the atomic coordinates (i.e. the crystal structure) and then allowing the indices h,k,l (where h,k,l are the components of the vector $\vec{h}$) to range uniformly over the integers. More precisely, it was found that, although the probability distribution of a structure factor $F_{\vec{h}}$ depends upon the vector $\vec{h}$, there are only a finite number of different distributions for each space group. Two vectors $\vec{h}_1$ and $\vec{h}_2$, such that the probability distributions of the corresponding structure factors $F_{\vec{h}_1}$ and $F_{\vec{h}_2}$ are identical, are called equivalent. In this way, for each space group, the totality of vectors $\vec{h}$ is subdivided into a finite number of equivalence classes, and any two vectors in any class are equivalent while no vector in any class is equivalent to any one in a different class. It turns out that if no relationship of the form $m_1x + m_2y + m_3z = m$ exists, where the m's are integers not all of which are zero (i.e. x,y,z are rationally independent) and x,y,z are the coordinates of any atom not in a fixed special position, then the probability distribution of an individual $F_{\vec{h}}$ (obtained by allowing the atoms in the asymmetric unit to range uniformly and independently throughout the unit cell) coincides with that obtained for $F_{\vec{h}_i}$ by fixing the atomic coordinates and allowing the vectors $\vec{h}_i$ to range uniformly through the vectors of the class to which $\vec{h}$ belongs. This is ultimately a consequence of the following facts. First, if h, k, and l are integers and x,y,z are uniformly and independently distributed in the interval (0,1), then the fractional part of hx + ky + lz is also uniformly distributed in the interval (0,1) (Appendix). Second, if x,y,z are rationally independent and h,k,l range uniformly over the integers, then the fractional part of hx + ky + lz is uniformly distributed in the interval (0,1) (Weyl, 1915-16). In short, the fractional part of hx + ky + lz is uniformly distributed whether h,k,l are fixed integers and x,y,z range uniformly and independently throughout the unit interval, or x,y,z are fixed, rationally independent numbers and h,k,l range uniformly and independently over all the integers. Finally, the structure factor is a function of the fractional part of hx + ky + lz. We conclude that the frequency distribution of the magnitudes of the structure factors obtained from experiment may be derived from our theoretical considerations for any crystal structure no matter how the atoms are arranged in the unit cell (subject to the rational independence of the coordinates previously mentioned and the measurement of a sufficient number of intensities).

In the third paper (Hauptman and Karle, 1953) probability distributions were obtained for the magnitude of a structure factor for every non-centrosymmetric space group. The most significant theoretical development was the introduction of the joint probability distribution. The joint distribution not only leads to probability distributions for both the magnitude and the phase of a structure factor, but also permits one to take into account an a priori knowledge of observed x-ray intensities. From a practical point of view, the introduction of the mixed moments was most significant since, with their aid, it was possible to evaluate in simple terms the complicated integrals which constantly appeared. Hence, the groundwork of the present theory was laid, and the program for solving the phase problem clearly indicated. In this Monograph the details of this program are carried out for all centrosymmetric space groups, and it is shown how the joint distributions lead to a routine procedure for sign determination.

In order to apply the final results of the present work, it is sufficient to become familiar with some of the definitions and with the simple formulas and procedures described in detail in the latter part of the Monograph. However, in order to follow the detailed mathematical arguments, it is desirable to become thoroughly acquainted with the mathematical theory developed in the earlier parts of the Monograph.

It is important to bear in mind that the observed intensities must be corrected for vibrational motion and placed on an absolute scale. If the data are sufficiently extensive this may be done by methods originally described by Wilson (1949) and discussed further in the Monograph.

We wish to thank Drs. J. D. H. Donnay, D. Harker, V. Luzzati, B. Magdoff, and A. L. Patterson for reading the original manuscript and for their many helpful discussions and comments. The Monograph has benefited materially therefrom. We are especially indebted to Dr. Donnay who burned the midnight oil to send us an extensive list of suggestions, and who devoted considerable time and effort to expedite publication of the Monograph.

Mr. Peter O'Hara of the Computation Laboratory of the National Bureau of Standards performed the required I.B.M. computations. His immediate understanding of the computational problems and facility in applying the computing techniques have been a great help to us, and his cooperation is gratefully acknowledged.

Herbert Hauptman
Jerome Karle

Washington, D.C.
June, 1953

CONTENTS

Contents (Continued)

Chapter 1

INTRODUCTION

Introductory Remarks

The underlying logical structure of the present Monograph may be briefly described in simple terms as follows. In the absence of any information concerning the values of the observable x-ray intensities, it is assumed that all positions of the atoms in the asymmetric unit of the unit cell are equally probable. In other words the atomic coordinates are assumed to be uniformly and independently distributed subject to the conditions imposed by symmetry. On this basis it is possible to determine the (*a priori*) probability distribution of a particular structure factor (Eq. (1.25) below). If now an experiment is performed and a certain set of x-ray intensities is observed, the probability distribution of the structure factor is changed, and we obtain the *a posteriori* probability distribution. Our basic postulate is that, provided the number of observed x-ray intensities is sufficiently large, the *a posteriori* distribution is essentially determined by the values of the observed intensities, and is relatively insensitive to the assumed *a priori* form of the atomic distribution (here taken to be uniform). For the purpose of this Monograph we need to assume even less; merely that the sign of a structure factor, as indicated by its *a posteriori* distribution, depends only on the values of a sufficiently large number of observed x-ray intensities. This assumption is plausible since it is well known that the crystal structure is determined (except for the possible existence of Patterson's homometric structures) by a sufficiently large number of x-ray intensities. It is therefore reasonable to suppose, provided the number of available intensities is greater than the minimum number required to determine the structure in the strict algebraic sense, that the sign of any structure factor as demanded by its *a posteriori* distribution will agree with its true sign (as required by the structure). It is clear that the greater the number of observed intensities, the greater will be the statistical significance of our procedure, and the more reliable the final answer.

The probability distribution $P_1(A)$ of a structure factor for any centrosymmetric crystal is an even function of A provided that the atoms in the asymmetric unit of the crystal are assumed to occupy all positions with equal probability (Eq. (1.25) below). Thus, the structure factor is just as likely to be positive as negative, even though its magnitude may be known. However, once a set of x-ray intensities is known, the atoms in a crystal no longer occupy all positions with equal probability. If the atoms in the asymmetric unit are assumed to range at random throughout the unit cell subject to the constraints imposed by the knowledge of a set of intensities, the resulting probability distribution of a structure factor is no longer an even function. The probability that the structure factor have a particular sign is now different from one-half. The purpose of this Monograph is to derive these probabilities on the basis that certain sets of intensities are specified and to derive therefrom a procedure for phase determination. Although the detailed mathematical analysis of the phase problem is quite complex, the final formulas for phase determination are relatively simple and can be applied in a routine fashion. The analogous problem for non-centrosymmetric crystals will be treated at a later date.

As is well known, the structure factor equations may be interpreted as coupled, plane, closed, vector polygons (Fig. 1). Consequently, from another point of view, this Monograph may be regarded as a study of the exact nature of the interdependence of these polygons. The solution of this problem is ultimately made to depend upon the evaluation of certain simple definite integrals, the mixed moments, to be defined later (Eq. (3.10)).

Review

In this section the probability distribution of a structure factor is derived for all centrosymmetric crystals (containing atoms in general positions only) on the basis that

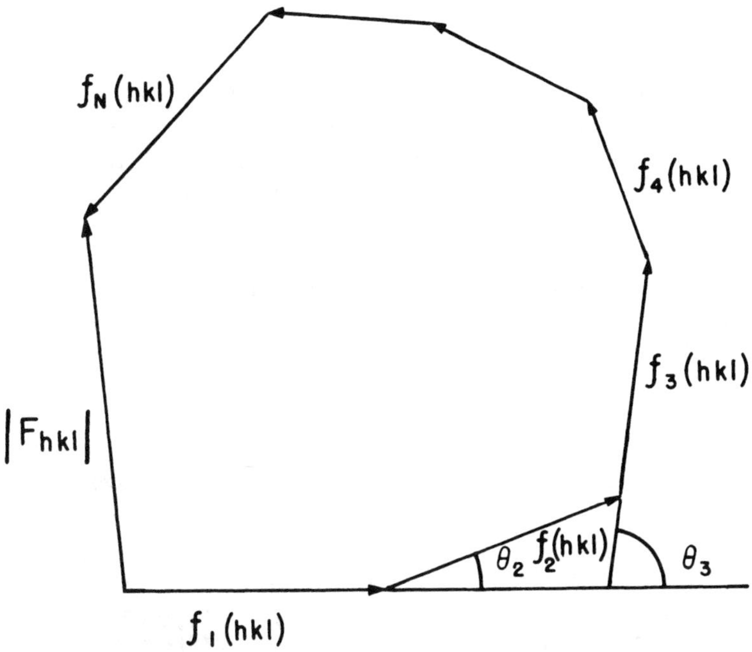

Fig. 1. Vector polygon interpretation of structure factor equation. Direction of f_1(hkl) is arbitrary and is taken as reference. $\theta_j = -2\pi(hx_j + ky_j + lz_j)$.

the atoms in the asymmetric unit range uniformly and independently throughout the unit cell. Wilson (1949) has already obtained approximate formulas for these distributions. However, we follow the procedure given in a previous paper (Karle and Hauptman, 1953a) in order to illustrate the mathematical methods which are generalized in this Monograph and which are needed for obtaining the more accurate formulas required.

The structure factor for the centrosymmetric crystal is given by

$$F_{\vec{h}} = \sum_{j=1}^{N/n} f_{j\vec{h}}\, \xi(x_j, y_j, z_j, \vec{h}), \qquad (1.01)$$

where n is the symmetry number or order of the space group, $f_{j\vec{h}}$ is the atomic scattering factor, N is the number of atoms in the unit cell, and $\xi(x_j, y_j, z_j, \vec{h})$ is some known trigonometric function of $\vec{h} = (h,k,l)$ and the atomic coordinates which depends on the space group, e.g. $\xi(x_j, y_j, z_j, \vec{h}) = 2\cos 2\pi(hx_j + ky_j + lz_j)$ for space group $P\bar{1}$ (International Tables for X-Ray Crystallography, Vol. I, 1952). The probability that $\xi_j = \xi(x_j, y_j, z_j, \vec{h})$ lie between c and c + dc is denoted by p(c)dc, where p(c) is a function which can be derived on the basis that the coordinates of the jth atom in the asymmetric unit are uniformly distributed. We make use of the property that p(c) is an even function which vanishes when the magnitude of c exceeds some positive number, b. Denote by Q(c) the probability that F be less than c. Then

$$Q(c) = \int \cdots \int_R p(\xi_1) \cdots p(\xi_{N/n}) d\xi_1 \cdots d\xi_{N/n}, \qquad (1.02)$$

where the region R in the ξ-space of N/n dimensions consists of all points $(\xi_1, \cdots, \xi_{N/n})$ such that the corresponding points $(x_j, y_j, z_j), j = 1, 2, \cdots, N/n$ yield by means of Eq. 1.01 a value of F which is less than c. Due to the very complex nature of the region R, Eq. 1.02 is conveniently replaced by

$$Q(c) = \int_{-\infty}^{\infty} \cdots \int_{-\infty}^{\infty} p(\xi_1) \cdots p(\xi_{N/n}) \times$$

$$T(\xi_1, \cdots, \xi_{N/n}) d\xi_1 \cdots d\xi_{N/n}, \qquad (1.03)$$

where

$$T(\xi_1, \cdots, \xi_{N/n}) = 0 \quad \text{if} \quad F > c, \qquad (1.04)$$

$$T(\xi_1, \cdots, \xi_{N/n}) = 1 \quad \text{if} \quad F < c. \qquad (1.05)$$

Evidently the introduction of the discontinuous function $T(\xi_1, \cdots, \xi_{N/n})$ enables us to replace the complicated region of integration R required by (1.02) by the much simpler region of (1.03), i.e. all of ξ-space. We choose for T the discontinuous integral

$$T(\xi_1, \cdots, \xi_{N/n}) = \frac{1}{2} - \frac{1}{\pi} \int_0^{\infty} \frac{\sin[(F - c)x]}{x} dx, \qquad (1.06)$$

so that (1.04) and (1.05) are satisfied. Then

$$Q(c) = \int_{-\infty}^{\infty} \cdots \int_{-\infty}^{\infty} \left(\frac{1}{2} - \frac{1}{\pi} \int_0^{\infty} \frac{\sin[(F - c)x]}{x} dx \right) \times$$

$$p(\xi_1) \cdots p(\xi_{N/n}) d\xi_1 \cdots d\xi_{N/n} \qquad (1.07)$$

$$= \frac{1}{2} - \frac{1}{\pi} \int_0^{\infty} \frac{dx}{x} \int_{-\infty}^{\infty} \cdots \int_{-\infty}^{\infty} p(\xi_1) \cdots p(\xi_{N/n}) \times$$

$$\sin[(F - c)x] d\xi_1 \cdots d\xi_{N/n}. \qquad (1.08)$$

We define

$$A_0 = 0, \quad A_k = \sum_{j=1}^{k} f_j \xi_j, \quad k = 1,2, \cdots, N/n. \qquad (1.09)$$

Then

$$A_{N/n} = F, \quad A_k = A_{k-1} + f_k \xi_k, \quad k = 1,2, \cdots, N/n, \qquad (1.10)$$

and

$$\int_{-\infty}^{\infty} p(\xi_k) \sin[(A_k - c)x] d\xi_k$$

$$= \int_{-\infty}^{\infty} p(\xi_k) \sin[(A_{k-1} + f_k \xi_k - c)x] d\xi_k \qquad (1.11)$$

$$= \int_{-\infty}^{\infty} p(\xi_k) \sin[(A_{k-1} - c)x] \cos(f_k \xi_k x) d\xi_k \qquad (1.12)$$

$$= \sin[(A_{k-1} - c)x] q(f_k x), \qquad (1.13)$$

where

$$q(f_k x) = \int_{-\infty}^{\infty} p(c) \cos(f_k x c) dc = 2 \int_0^{\infty} p(c) \cos(f_k x c) dc. \qquad (1.14)$$

Repeated application of (1.13) enables us to replace (1.08) by

$$Q(c) = \frac{1}{2} + \frac{1}{\pi} \int_0^{\infty} \frac{\sin(cx)}{x} \prod_{k=1}^{N/n} q(f_k x) dx. \qquad (1.15)$$

Denote by $P_1(A)dA$ the probability that A be between A and A + dA. In view of

$$P_1(A) = \frac{d}{dA} Q(A), \tag{1.16}$$

(1.15) implies

$$P_1(A) = \frac{1}{\pi}\int_0^\infty \cos Ax \prod_{k=1}^{N/n} q(f_k x)dx. \tag{1.17}$$

From (1.14), $q(f_k x)$ may be expressed in the form of a series

$$q(f_k x) = 1 - \frac{f_k^2}{2!} m_2 x^2 + \frac{f_k^4}{4!} m_4 x^4 - \cdots, \tag{1.18}$$

where

$$m_j = \int_{-\infty}^\infty c^j p(c)dc \tag{1.19}$$

is the jth moment of p(c). Consequently, in order to find $q(f_k x)$ it is not necessary to determine the function p(c) explicitly. Instead, we need only determine the jth moment of p(c) for even values of j. Evidently the jth moment of p(c) is the expected (or average) value of ξ^j and is therefore given by

$$m_j = \int_0^1\int_0^1\int_0^1 \xi^j(x, y, z)\,dx\,dy\,dz. \tag{1.20}$$

From (1.18), using the Maclaurin expansion of the logarithm,

$$\log q(f_k x) = -\frac{f_k^2}{2!} m_2 x^2 - \frac{f_k^4}{4!}(3m_2^2 - m_4)x^4$$

$$- \frac{f_k^6}{6!}(30m_2^3 - 15m_2 m_4 + m_6)x^6 - \cdots, \tag{1.21}$$

and

$$\log \prod_{k=1}^{N/n} q(f_k x) = -\frac{\sigma_2}{2!n} m_2 x^2 - \frac{\sigma_4}{4!n}(3m_2^2 - m_4)x^4$$

$$- \frac{\sigma_6}{6!n}(30m_2^3 - 15m_2 m_4 + m_6)x^6 - \cdots, \tag{1.22}$$

where

$$\sigma_j = \sum_{k=1}^{N} f_k^j = n \sum_{k=1}^{N/n} f_k^j. \tag{1.23}$$

Hence

$$\prod_{k=1}^{N/n} q(f_k x) = \exp\left(-\frac{\sigma_2}{2n} m_2 x^2\right) \left\{1 - \frac{\sigma_4}{4!n}(3m_2^2 - m_4)x^4 \right.$$

$$\left. - \frac{\sigma_6}{6!n}(30m_2^3 - 15m_2m_4 + m_6)x^6 - \cdots\right\}. \tag{1.24}$$

Substituting into (1.17) we find

$$P_1(A) = \sqrt{\frac{n}{2\pi m_2\sigma_2}} \exp\left(-\frac{A^2 n}{2m_2\sigma_2}\right)\left\{1 - \frac{n\sigma_4(3m_2^2 - m_4)}{2\cdot 4m_2^2\sigma_2^2} \times \right.$$

$$\left(1 - \frac{2nA^2}{m_2\sigma_2} + \frac{n^2A^4}{3m_2^2\sigma_2^2}\right) - \frac{n^2\sigma_6(30m_2^3 - 15m_2m_4 + m_6)}{2\cdot 4\cdot 6m_2^3\sigma_2^3} \times$$

$$\left.\left(1 - \frac{3nA^2}{m_2\sigma_2} + \frac{n^2A^4}{m_2^2\sigma_2^2} - \frac{n^3A^6}{15m_2^3\sigma_2^3}\right) - \cdots\right\}. \tag{1.25}$$

From (1.25) the average value of any power of F may be obtained immediately, e.g.

$$\langle F^2\rangle = \int_{-\infty}^{\infty} A^2 P_1(A)dA = \frac{m_2}{n}\sigma_2. \tag{1.26}$$

If we define the <u>normalized structure factor</u> E by means of

$$E = \frac{F}{\frac{m_2}{n}\sigma_2}, \tag{1.27}$$

(1.26) implies

$$\langle E^2\rangle = 1. \tag{1.28}$$

In terms of the normalized structure factor the probability distributions assume a very simple form. Denote by $P(E)dE$ the probability that E lie between E and E + dE.

Then from (1.25) we have

$$P(E) = \frac{e^{-\frac{1}{2}E^2}}{\sqrt{2\pi}} \Bigg\{ 1 -$$

$$\frac{n\sigma_4(3m_2^2 - m_4)}{2 \cdot 4m_2^2\sigma_2^2} \left(1 - 2E^2 + \frac{1}{3}E^4\right)$$

$$- \frac{n^2\sigma_6(30m_2^3 - 15m_2m_4 + m_6)}{2 \cdot 4 \cdot 6m_2^3\sigma_2^3} \left(1 - 3E^2 + E^4 - \frac{1}{15}E^6\right) - \cdots \Bigg\}. \tag{1.29}$$

The form of (1.29) is independent of the space group; only the numerical coefficients of the polynomials depend on the particular space group (as well as on the indices h,k,l).

In the case that the crystal contains atoms in special positions as well as in general positions, analogous distributions may be readily found (Karle and Hauptman, 1953 a).

Chapter 2

PRELIMINARIES

Formulation

Since the joint probability distribution of the structure factors leads to a practical solution of the phase problem, it is appropriate to formulate this problem in a precise fashion. The phase problem is the problem of determining the phases (either 0 or π) of the structure factors $F_{\vec{h}}$, defined by (1.01), given the magnitudes of $F_{\vec{h}}$ and the values of $f_{j\vec{h}}$ for a sufficiently large number of vectors $\vec{h}$.

The crystal structure alone does not however determine all the phases, because (1.01) implies that an appropriate origin has been selected. In fact, both the functional form of ξ and the values of the atomic coordinates x_j, y_j, z_j depend upon which center of symmetry is chosen as origin. The magnitude of $F_{\vec{h}}$ is, of course, independent of the choice of origin. Since in general a finite number of origins is permissible, the phase of a structure factor depends not only on the structure but also on the choice of origin. However, as will be seen, there always exist certain linear combinations of the phases whose values (reduced modulo 2π) depend upon the structure alone and are independent of the choice of origin, and therefore of the form of the structure factor also. We shall call them <u>structure invariants</u>. Furthermore, <u>for a fixed functional form of the structure factor</u>, there always exist certain linear combinations of the phases whose values (reduced modulo 2π) depend upon the structure alone and are independent of the choice of origin permitted by the chosen form of the structure factor. We shall call them <u>structure seminvariants</u>. It turns out that the structure seminvariants are independent of the chosen fixed functional form of the structure factor. While the structure seminvariants depend only on the space group and the choice of unit cell, their <u>values</u>, for a given crystal, depend on the chosen form of the structure factor. Evidently every structure invariant is also a structure seminvariant. In fact, those structure seminvariants whose values are independent of the chosen functional form of the structure factor coincide with the structure invariants.

Since the knowledge of a sufficient number of intensities determines the crystal structure, these intensities determine also the values of the structure invariants rather than the phases themselves. Therefore the phase problem may be more accurately described as the problem of determining the values of the structure invariants once a sufficiently large number of intensities has been given (assuming as always that the atomic structure factors are known). From another point of view the phase problem is the problem of determining the values of the structure seminvariants, for each fixed form of structure factor, once a sufficiently large number of intensities is known. The phases may then be obtained from the values of the structure seminvariants by specifying the origin. It is seen that in any solution of the phase problem, an essential requirement is the identification of the structure invariants and seminvariants for each space group. It should be emphasized that the probability theory will afford a solution of this problem. However an a priori knowledge of both the structure seminvariants and the structure invariants is an invaluable aid in interpreting the results of the probability theory, and the following sections are devoted to a discussion of this subject for crystals having atoms in general positions only.

Linear Dependence and Independence Modulo Two

First the concept of linear dependence modulo 2 is introduced since it permits the statement of general conclusions in a convenient and concise fashion. As is well known, if the integer h is even, h is said to be congruent to zero modulo two, and we write

$$h \equiv 0 \pmod 2). \tag{2.01}$$

Similarly we say that the vector $\vec{h} = (h_1, h_2, \cdots, h_p)$, where $h_1, h_2, \cdots, h_p$ are integers, is even or congruent to $0 \pmod 2$ if each of $h_1, h_2 \cdots, h_p$ is even, and we write

$$\vec{h} \equiv 0 \pmod{2}. \tag{2.02}$$

Two vectors $\vec{h}_1$ and $\vec{h}_2$ are congruent modulo 2 if the difference $\vec{h}_1 - \vec{h}_2$ is even; and the notation

$$\vec{h}_1 \equiv \vec{h}_2 \pmod{2} \tag{2.03}$$

is used. A set of n vectors $\vec{h}_j$, $j = 1, 2, \cdots, n$, $(n \geqslant 1)$, is said to be linearly dependent modulo 2 if there exists a set of n integers $a_j = 0$ or 1, $j = 1, 2, \cdots, n$, not all of which are zero, such that

$$\sum_{j=1}^{n} a_j \vec{h}_j \equiv 0 \pmod{2}. \tag{2.04}$$

Otherwise the set $\vec{h}_j$ is said to be linearly independent modulo 2. Finally the vector $\vec{h}$ is linearly dependent modulo 2 on, or linearly independent modulo 2 of, the set $\vec{h}_j$, $j = 1, 2, \cdots, n$, $(n \geqslant 1)$, according as there exist or there do not exist n integers $a_j = 0$ or 1 such that

$$\vec{h} \equiv \sum_{j=1}^{n} a_j \vec{h}_j \pmod{2}. \tag{2.05}$$

For example $(24\bar{2})$ is linearly dependent modulo 2. Any vector $\vec{h}$ which is linearly dependent modulo 2 (i.e. congruent to zero modulo 2) is also linearly dependent modulo 2 on any set of vectors $\vec{h}_j$, $j = 1, 2, \cdots, n$, as may be seen by taking every a_j in (2.05) to be zero. (386) and (124) are congruent modulo 2, since (262) is even. Again, (124) is linearly dependent modulo 2 on (346) but is linearly independent modulo 2 of the pair (234), (465). Evidently (100), (010), (001) constitute a set of three vectors which is linearly independent modulo 2. Any set of four vectors however, each having three components, is linearly dependent modulo 2.

Equivalence and Similarity

In general the function ξ which defines the structure factor (1.01) depends upon which of the centers of symmetry is chosen as origin. Two origins will be called equivalent if the functional forms of ξ to which they give rise are identical. In other words, two origins are equivalent if they are geometrically related in the same way to all the symmetry elements. It is important to distinguish between the value of ξ and the form of ξ. The values of ξ referred to two equivalent origins may be equal or may be the negatives of each other. Likewise their values referred to two non-equivalent origins may be equal or negatives of each other. The concept of equivalent origins leads to the notion of equivalence classes. The set of all origins (centers of symmetry) may be grouped into classes, and any two origins in any class are equivalent while no origin in one class is equivalent to any origin in a different class.

The phase of a structure factor depends in general upon the choice of origin. Two centers (origins) will be called similar if every structure seminvariant has the same value when referred to either center as origin. If two origins are equivalent then they are obviously similar. Furthermore, it will be seen later that the converse is also true.* The concept of similar origins leads to the notion of similarity classes. The set of all centers of symmetry may be grouped into classes, and any two centers in any class are similar while no center in one class is similar to any center in a different class.

As previously pointed out, the values of only the structure invariants are determined by the structure, while the values of the phases depend also on which center is chosen as origin. It will be seen that the origin may be chosen by first selecting the form of the structure factor, i.e. an equivalence class, and then by specifying arbitrarily the signs of an appropriate set of structure factors.

The Four Types of Centrosymmetric Space Groups

A survey of the 92 centrosymmetric space groups shows that for primitive unit cells there are always eight permissible origins (permissible in that they are centers of symmetry). Our next concern is how the value of the structure factor changes as we move from one such origin to another. Although the probability theory is valid regardless of the

*provided that the unit cell is chosen to be primitive.

choice of the unit cell, this preliminary discussion is restricted to primitive cells so that it may be kept within reasonable bounds. In "International Tables for X-Ray Crystallography", Vol. I, 1952, the unit cell is chosen to be primitive for 62 of the 92 centrosymmetric space groups. However, for the remaining 30 space groups, structure factors appropriate to the choice of a primitive unit cell are readily obtained. Once this is done the results of this discussion become applicable to these space groups also. In the monoclinic system we have chosen the second setting with b-axis unique.

The structure factor for the general centrosymmetric crystal having N atoms per unit cell may be written

$$F_{hkl} = 2 \sum_{j=1}^{N/2} f_j \cos 2\pi(hx_j + ky_j + lz_j) \qquad (2.06)$$

where the atomic structure factor of the jth atom, f_j, is a function of h,k,l, and the coordinates of the jth atom are x_j, y_j, z_j. If the origin is shifted to a different center having coordinates ϵ_1, ϵ_2, ϵ_3 with respect to the first origin, then x_j, y_j, z_j in (2.06) is replaced by $x_j - \epsilon_1$, $y_j - \epsilon_2$, $z_j - \epsilon_3$. Since the unit cell is primitive $\epsilon_i = 0$ or $1/2$, $i = 1, 2, 3$, and it is readily verified that (2.06) is replaced by

$$F'_{hkl} = 2 \sum_{j=1}^{N/2} f_j \cos 2\pi(hx_j + ky_j + lz_j) \times$$

$$\cos 2\pi(h\epsilon_1 + k\epsilon_2 + l\epsilon_3)$$

$$= F_{hkl} \cos 2\pi(h\epsilon_1 + k\epsilon_2 + l\epsilon_3). \qquad (2.07)$$

In short, F_{hkl} is multiplied by +1 or -1 according as $2(h\epsilon_1 + k\epsilon_2 + l\epsilon_3)$ is an even or odd integer.

The centrosymmetric space groups fall into three different categories depending upon the number of equivalence classes. Category 1 consists of those space groups having one equivalence class, Category 2 of those space groups having two equivalence classes, and Category 3 of those having four equivalence classes. As shown in Table 1,

Table 1. The four types of centrosymmetric space group and, for each type and a fixed form of structure factor, the phases whose values are to be specified arbitrarily. The values of the remaining phases are then all determined.

Category	1			2	3		
No. of equivalence classes	1			2	4		
Type	1P			2P	$3P_1$	$3P_2$	
Crystal System	Triclinic	Monoclinic	Orthorhombic	Tetragonal	Hexagonal	Rhombohedral	Cubic
Space groups	$P\bar{1}$	$P2/m$	Pmmm	$P4/m$	$P\bar{3}$	$R\bar{3}$	Pm3
		$P2_1/m$	Pnnn	$P4_2/m$	$P\bar{3}1m$	$R\bar{3}m$	Pn3
		$P2/c$	Pccm	$P4/n$	$P\bar{3}1c$	$R\bar{3}c$	Pa3
		$P2_1/c$	Pban	$P4_2/n$	$P\bar{3}m1$		Pm3m
			Pmma	$P4/mmm$	$P\bar{3}c1$		Pn3n
			Pnna	$P4/mcc$	$P6/m$		Pm3n
			Pmna	$P4/nbm$	$P6_3/m$		Pn3m
			Pcca	$P4/nnc$	$P6/mmm$		
			Pbam	$P4/mbm$	$P6/mcc$		
			Pccn	$P4/mnc$	$P6_3/mcm$		
			Pbcm	$P4/nmm$	$P6_3/mmc$		
			Pnnm	$P4/ncc$			
			Pmmn	$P4_2/mmc$			
			Pbcn	$P4_2/mcm$			
			Pbca	$P4_2/nbc$			
			Pnma	$P4_2/nnm$			
				$P4_2/mbc$			
				$P4_2/mnm$			
				$P4_2/nmc$			
				$P4_2/ncm$			

Permissible origins (centers)	0, 0, 0 $\frac{1}{2}$, 0, 0 0, $\frac{1}{2}$, 0 0, 0, $\frac{1}{2}$ 0, $\frac{1}{2}$, $\frac{1}{2}$ $\frac{1}{2}$, 0, $\frac{1}{2}$ $\frac{1}{2}$, $\frac{1}{2}$, 0 $\frac{1}{2}$, $\frac{1}{2}$, $\frac{1}{2}$	[0, 0, 0 0, 0, $\frac{1}{2}$ $\frac{1}{2}$, $\frac{1}{2}$, 0 $\frac{1}{2}$, $\frac{1}{2}$, $\frac{1}{2}$] [$\frac{1}{2}$, 0, 0 0, $\frac{1}{2}$, 0 0, $\frac{1}{2}$, $\frac{1}{2}$ $\frac{1}{2}$, 0, $\frac{1}{2}$]	[0, 0, 0 0, 0, $\frac{1}{2}$] [$\frac{1}{2}$, 0, 0 $\frac{1}{2}$, 0, $\frac{1}{2}$] [0, $\frac{1}{2}$, 0 0, $\frac{1}{2}$, $\frac{1}{2}$] [$\frac{1}{2}$, $\frac{1}{2}$, 0 $\frac{1}{2}$, $\frac{1}{2}$, $\frac{1}{2}$]	[0, 0, 0 $\frac{1}{2}$, $\frac{1}{2}$, $\frac{1}{2}$] [$\frac{1}{2}$, 0, 0 0, $\frac{1}{2}$, $\frac{1}{2}$] [0, $\frac{1}{2}$, 0 $\frac{1}{2}$, 0, $\frac{1}{2}$] [0, 0, $\frac{1}{2}$ $\frac{1}{2}$, $\frac{1}{2}$, 0]
Vector $\vec{h}_i$ invariantly associated with $\phi_{\vec{h}}$ or with $\vec{h}$ = (h,k,l)	(h,k,l)	(h,k,l)	(h,k,l)	(h,k,l)
Vector $\vec{h}_s$ seminvariantly associated with $\phi_{\vec{h}}$ or with $\vec{h}$ = (h,k,l)	(h,k,l)	(h+k,l)	(l)	(h+k+l)
No. of phases, lin. semi-ind. mod 2, to be specified arbitrarily	3	2	1	

Category 3 is further subdivided into two types depending upon the nature of the equivalence classes. Each type is clearly characterized by row 6 of Table 1, the equivalence classes being defined by the boxes with solid lines.

Type 1P (Category 1). For each space group there is a unique form for the structure factor. Hence there is only one equivalence class and one similarity class. For each of the eight possible sets of indices heading the eight columns of Table 2, we group in the same box all centers referred to which the corresponding structure factor has the same sign. From (2.07), it is seen that the structure factor will have the same sign when referred to any of the eight permissible origins provided that $h\epsilon_1 + k\epsilon_2 + l\epsilon_3$ is an integer for all eight points $\epsilon_1, \epsilon_2, \epsilon_3$, including $\epsilon_1 = \epsilon_2 = \epsilon_3 = 0$, i.e., if and only if $\vec{h} \equiv 0 (\mathrm{mod}\ 2)$. Hence all eight centers in the first column of Table 2 are included in the same box. In the second column, $h_1 \not\equiv 0\ (\mathrm{mod}\ 2)$ while $k_1 \equiv l_1 \equiv 0\ (\mathrm{mod}\ 2)$. It is readily verified that any structure factor $F_{\vec{h}_1}$, where $\vec{h}_1$ satisfies these conditions, has the same value when referred to each of the four centers in the upper box, but its sign is reversed when referred to any of the four centers in the lower box. Similar remarks apply to columns 3 to 8 of Table 2, in whose headings the subscript of $\vec{h}$ indicates which indices are odd.

The problem which remains can be most easily clarified by selecting a particular set of three vectors which is linearly independent modulo 2, e.g., $\vec{h}_j$, $j = 1, 2, 3$, where

$$h_1 \not\equiv 0,\ k_1 \equiv l_1 \equiv 0\ (\mathrm{mod}\ 2),$$

$$h_2 \equiv 0,\ k_2 \not\equiv 0, l_2 \equiv 0\ (\mathrm{mod}\ 2),$$

$$h_3 \equiv k_3 \equiv 0, l_3 \not\equiv 0\ (\mathrm{mod}\ 2).$$

Then the relationship among columns 2, 3, and 5 of Table 2, where $\vec{h}_{12}$ is linearly dependent modulo 2 on the pair $\vec{h}_1$, $\vec{h}_2$, may be described by observing that those centers which lie both in a box of column 2 and a box of column 3 also lie in a single box of column 5. Similar remarks apply to columns 2, 4, and 6 and to columns 3, 4, and 7. Finally one and only one center lies in each of a box of column 2, a box of column 3, and a box of column 4. Similar conclusions are obtained no matter which set of three vectors, which is

Table 2. Sets of centers which give rise to the same value of the structure factor for each of the eight possible sets of indices for space groups of Type 1P.

$\vec{h} \equiv 0 \pmod 2$	$\vec{h}_1$	$\vec{h}_2$	$\vec{h}_3$	$\vec{h}_{12}$	$\vec{h}_{13}$	$\vec{h}_{23}$	$\vec{h}_{123}$
h even	h_1 odd	h_2 even	h_3 even	h_{12} odd	h_{13} odd	h_{23} even	h_{123} odd
k even	k_1 even	k_2 odd	k_3 even	k_{12} odd	k_{13} even	k_{23} odd	k_{123} odd
l even	l_1 even	l_2 even	l_3 odd	l_{12} even	l_{13} odd	l_{23} odd	l_{123} odd
$0, 0, 0$	$0, 0, 0$	$0, 0, 0$	$0, 0, 0$	$0, 0, 0$	$0, 0, 0$	$0, 0, 0$	$0, 0, 0$
$0, 0, \frac{1}{2}$	$0, 0, \frac{1}{2}$	$0, 0, \frac{1}{2}$	$0, \frac{1}{2}, 0$	$0, 0, \frac{1}{2}$	$0, \frac{1}{2}, 0$	$\frac{1}{2}, 0, 0$	$0, \frac{1}{2}, \frac{1}{2}$
$0, \frac{1}{2}, 0$	$0, \frac{1}{2}, 0$	$\frac{1}{2}, 0, 0$	$\frac{1}{2}, 0, 0$	$\frac{1}{2}, \frac{1}{2}, 0$	$\frac{1}{2}, 0, \frac{1}{2}$	$0, \frac{1}{2}, \frac{1}{2}$	$\frac{1}{2}, 0, \frac{1}{2}$
$\frac{1}{2}, 0, 0$	$0, \frac{1}{2}, \frac{1}{2}$	$\frac{1}{2}, 0, \frac{1}{2}$	$\frac{1}{2}, \frac{1}{2}, 0$	$\frac{1}{2}, \frac{1}{2}, \frac{1}{2}$	$\frac{1}{2}, \frac{1}{2}, \frac{1}{2}$	$\frac{1}{2}, \frac{1}{2}, \frac{1}{2}$	$\frac{1}{2}, \frac{1}{2}, 0$
$0, \frac{1}{2}, \frac{1}{2}$	$\frac{1}{2}, 0, 0$	$0, \frac{1}{2}, 0$	$0, 0, \frac{1}{2}$	$0, \frac{1}{2}, 0$	$0, 0, \frac{1}{2}$	$0, 0, \frac{1}{2}$	$0, 0, \frac{1}{2}$
$\frac{1}{2}, 0, \frac{1}{2}$	$\frac{1}{2}, 0, \frac{1}{2}$	$0, \frac{1}{2}, \frac{1}{2}$	$0, \frac{1}{2}, \frac{1}{2}$	$0, \frac{1}{2}, \frac{1}{2}$	$0, \frac{1}{2}, \frac{1}{2}$	$\frac{1}{2}, 0, \frac{1}{2}$	$0, \frac{1}{2}, 0$
$\frac{1}{2}, \frac{1}{2}, 0$	$\frac{1}{2}, \frac{1}{2}, 0$	$\frac{1}{2}, \frac{1}{2}, 0$	$\frac{1}{2}, 0, \frac{1}{2}$	$\frac{1}{2}, 0, 0$	$\frac{1}{2}, 0, 0$	$0, \frac{1}{2}, 0$	$\frac{1}{2}, 0, 0$
$\frac{1}{2}, \frac{1}{2}, \frac{1}{2}$	$\frac{1}{2}, \frac{1}{2}, \frac{1}{2}$	$\frac{1}{2}, \frac{1}{2}, \frac{1}{2}$	$\frac{1}{2}, \frac{1}{2}, \frac{1}{2}$	$\frac{1}{2}, 0, \frac{1}{2}$	$\frac{1}{2}, \frac{1}{2}, 0$	$\frac{1}{2}, \frac{1}{2}, 0$	$\frac{1}{2}, \frac{1}{2}, \frac{1}{2}$

Table 3. Sets of centers which give rise to the same value of the structure factor for each of the eight possible sets of indices for space groups of Type 2P. Broken lines separate non-equivalent origins.

$\vec{h}$	$\vec{h}_1$	$\vec{h}_2$	$\vec{h}_{12}$	$\vec{h}_3$	$\vec{h}_{13}$	$\vec{h}_{23}$	$\vec{h}_{123}$
h even	h_1 odd	h_2 odd	h_{12} even	h_3 even	h_{13} odd	h_{23} odd	h_{123} even
k even	k_1 odd	k_2 even	k_{12} odd	k_3 even	k_{13} odd	k_{23} even	k_{123} odd
l even	l_1 even	l_2 even	l_{12} even	l_3 odd	l_{13} odd	l_{23} odd	l_{123} odd
0, 0, 0	0, 0, 0	0, 0, 0	0, 0, 0	0, 0, 0	0, 0, 0	0, 0, 0	0, 0, 0
0, 0, $\frac{1}{2}$	0, 0, $\frac{1}{2}$	0, 0, $\frac{1}{2}$	0, 0, $\frac{1}{2}$	$\frac{1}{2}$, $\frac{1}{2}$, 0	$\frac{1}{2}$, $\frac{1}{2}$, 0	$\frac{1}{2}$, $\frac{1}{2}$, $\frac{1}{2}$	$\frac{1}{2}$, $\frac{1}{2}$, $\frac{1}{2}$
$\frac{1}{2}$, $\frac{1}{2}$, 0	$\frac{1}{2}$, $\frac{1}{2}$, 0	0, $\frac{1}{2}$, 0	$\frac{1}{2}$, 0, 0	0, $\frac{1}{2}$, 0	0, $\frac{1}{2}$, $\frac{1}{2}$	0, $\frac{1}{2}$, 0	$\frac{1}{2}$, 0, 0
$\frac{1}{2}$, $\frac{1}{2}$, $\frac{1}{2}$	$\frac{1}{2}$, $\frac{1}{2}$, $\frac{1}{2}$	0, $\frac{1}{2}$, $\frac{1}{2}$	$\frac{1}{2}$, 0, $\frac{1}{2}$	$\frac{1}{2}$, 0, 0	$\frac{1}{2}$, 0, $\frac{1}{2}$	$\frac{1}{2}$, 0, $\frac{1}{2}$	0, $\frac{1}{2}$, $\frac{1}{2}$
0, $\frac{1}{2}$, 0	0, $\frac{1}{2}$, 0	$\frac{1}{2}$, $\frac{1}{2}$, 0	$\frac{1}{2}$, $\frac{1}{2}$, 0	0, 0, $\frac{1}{2}$	0, 0, $\frac{1}{2}$	0, 0, $\frac{1}{2}$	0, 0, $\frac{1}{2}$
$\frac{1}{2}$, 0, 0	$\frac{1}{2}$, 0, 0	$\frac{1}{2}$, $\frac{1}{2}$, $\frac{1}{2}$	$\frac{1}{2}$, $\frac{1}{2}$, $\frac{1}{2}$	$\frac{1}{2}$, $\frac{1}{2}$, $\frac{1}{2}$	$\frac{1}{2}$, $\frac{1}{2}$, $\frac{1}{2}$	$\frac{1}{2}$, $\frac{1}{2}$, 0	$\frac{1}{2}$, $\frac{1}{2}$, 0
0, $\frac{1}{2}$, $\frac{1}{2}$	0, $\frac{1}{2}$, $\frac{1}{2}$	$\frac{1}{2}$, 0, 0	0, $\frac{1}{2}$, 0	0, $\frac{1}{2}$, $\frac{1}{2}$	0, $\frac{1}{2}$, 0	$\frac{1}{2}$ 0, 0	0, $\frac{1}{2}$, 0
$\frac{1}{2}$, 0, $\frac{1}{2}$	$\frac{1}{2}$, 0, $\frac{1}{2}$	$\frac{1}{2}$, 0, $\frac{1}{2}$	0, $\frac{1}{2}$, $\frac{1}{2}$	$\frac{1}{2}$, 0, $\frac{1}{2}$	$\frac{1}{2}$, 0, 0	0, $\frac{1}{2}$, $\frac{1}{2}$	$\frac{1}{2}$, 0, $\frac{1}{2}$

Table 4. For each of the two equivalence classes, sets of centers which give rise to the same value of the structure factor for each of the four possible sets of indices for space groups of Type 2P.

<table>
<tr><td>$\vec{h}$</td><td>$\vec{h}_1$</td><td>$\vec{h}_2$</td><td>$\vec{h}_{12}$</td></tr>
<tr><td>$h+k$ even</td><td>h_1+k_1 odd</td><td>h_2+k_2 even</td><td>$h_{12}+k_{12}$ odd</td></tr>
<tr><td>l even</td><td>l_1 even</td><td>l_2 odd</td><td>l_{12} odd</td></tr>
<tr><td rowspan="2">$0, 0, 0$
$0, 0, \frac{1}{2}$
$\frac{1}{2}, \frac{1}{2}, 0$
$\frac{1}{2}, \frac{1}{2}, \frac{1}{2}$</td><td>$0, 0, 0$
$0, 0, \frac{1}{2}$</td><td>$0, 0, 0$
$\frac{1}{2}, \frac{1}{2}, 0$</td><td>$0, 0, 0$
$\frac{1}{2}, \frac{1}{2}, \frac{1}{2}$</td></tr>
<tr><td>$\frac{1}{2}, \frac{1}{2}, 0$
$\frac{1}{2}, \frac{1}{2}, \frac{1}{2}$</td><td>$0, 0, \frac{1}{2}$
$\frac{1}{2}, \frac{1}{2}, \frac{1}{2}$</td><td>$0, 0, \frac{1}{2}$
$\frac{1}{2}, \frac{1}{2}, 0$</td></tr>
<tr><td colspan="4"></td></tr>
<tr><td rowspan="2">$0, \frac{1}{2}, 0$
$\frac{1}{2}, 0, 0$
$0, \frac{1}{2}, \frac{1}{2}$
$\frac{1}{2}, 0, \frac{1}{2}$</td><td>$0, \frac{1}{2}, 0$
$0, \frac{1}{2}, \frac{1}{2}$</td><td>$0, \frac{1}{2}, 0$
$\frac{1}{2}, 0, 0$</td><td>$0, \frac{1}{2}, 0$
$\frac{1}{2}, 0, \frac{1}{2}$</td></tr>
<tr><td>$\frac{1}{2}, 0, 0$
$\frac{1}{2}, 0, \frac{1}{2}$</td><td>$0, \frac{1}{2}, \frac{1}{2}$
$\frac{1}{2}, 0, \frac{1}{2}$</td><td>$\frac{1}{2}, 0, 0$
$0, \frac{1}{2}, \frac{1}{2}$</td></tr>
</table>

Table 5. Sets of centers which give rise to the same value of the structure factor for each of the eight possible sets of indices for space groups of Type $3P_1$. Broken lines separate non-equivalent origins.

$\vec{h}$	$\vec{h}_1$	$\vec{h}_2$	$\vec{h}_{12}$	$\vec{h}_3$	$\vec{h}_{13}$	$\vec{h}_{23}$	$\vec{h}_{123}$
h even	h_1 even	h_2 odd	h_{12} odd	h_3 even	h_{13} even	h_{23} odd	h_{123} odd
k even	k_1 odd	k_2 even	k_{12} odd	k_3 even	k_{13} odd	k_{23} even	k_{123} odd
l even	l_1 even	l_2 even	l_{12} even	l_3 odd	l_{13} odd	l_{23} odd	l_{123} odd
0, 0, 0	0, 0, 0	0, 0, 0	0, 0, 0	0, 0, 0	0, 0, 0	0, 0, 0	0, 0, 0
0, 0, $\frac{1}{2}$	0, 0, $\frac{1}{2}$	0, 0, $\frac{1}{2}$	0, 0, $\frac{1}{2}$	0, $\frac{1}{2}$, 0	0, $\frac{1}{2}$, $\frac{1}{2}$	0, $\frac{1}{2}$, 0	0, $\frac{1}{2}$, $\frac{1}{2}$
0, $\frac{1}{2}$, 0	$\frac{1}{2}$, 0, 0	0, $\frac{1}{2}$, 0	$\frac{1}{2}$, $\frac{1}{2}$, 0	$\frac{1}{2}$, 0, 0	$\frac{1}{2}$, 0, 0	$\frac{1}{2}$, 0, $\frac{1}{2}$	$\frac{1}{2}$, 0, $\frac{1}{2}$
0, $\frac{1}{2}$, $\frac{1}{2}$	$\frac{1}{2}$, 0, $\frac{1}{2}$	0, $\frac{1}{2}$, $\frac{1}{2}$	$\frac{1}{2}$, $\frac{1}{2}$, $\frac{1}{2}$	$\frac{1}{2}$, $\frac{1}{2}$, 0	$\frac{1}{2}$, $\frac{1}{2}$, $\frac{1}{2}$,	$\frac{1}{2}$, $\frac{1}{2}$, $\frac{1}{2}$	$\frac{1}{2}$, $\frac{1}{2}$, 0
$\frac{1}{2}$, 0, 0	0, $\frac{1}{2}$, 0	$\frac{1}{2}$, 0, 0	0, $\frac{1}{2}$, 0	0, 0, $\frac{1}{2}$	0, 0, $\frac{1}{2}$	0, 0, $\frac{1}{2}$	0, 0, $\frac{1}{2}$
$\frac{1}{2}$, 0, $\frac{1}{2}$	0, $\frac{1}{2}$, $\frac{1}{2}$	$\frac{1}{2}$, 0, $\frac{1}{2}$	0, $\frac{1}{2}$, $\frac{1}{2}$	0, $\frac{1}{2}$, $\frac{1}{2}$	0, $\frac{1}{2}$, 0	0, $\frac{1}{2}$, $\frac{1}{2}$	0, $\frac{1}{2}$, 0
$\frac{1}{2}$, $\frac{1}{2}$, 0	$\frac{1}{2}$, $\frac{1}{2}$, 0	$\frac{1}{2}$, $\frac{1}{2}$, 0	$\frac{1}{2}$, 0, 0	$\frac{1}{2}$, 0, $\frac{1}{2}$	$\frac{1}{2}$, 0, $\frac{1}{2}$	$\frac{1}{2}$, 0, 0	$\frac{1}{2}$, 0, 0
$\frac{1}{2}$, $\frac{1}{2}$, $\frac{1}{2}$	$\frac{1}{2}$, $\frac{1}{2}$, $\frac{1}{2}$,	$\frac{1}{2}$, $\frac{1}{2}$, $\frac{1}{2}$	$\frac{1}{2}$, 0, $\frac{1}{2}$	$\frac{1}{2}$, $\frac{1}{2}$, $\frac{1}{2}$	$\frac{1}{2}$, $\frac{1}{2}$, 0	$\frac{1}{2}$, $\frac{1}{2}$, 0	$\frac{1}{2}$, $\frac{1}{2}$, $\frac{1}{2}$

Table 6. For each of the four equivalence classes, sets of centers which give rise to the same value of the structure factor for each of the two possible sets of indices for space groups of Type $3P_1$.

$\vec{h}$ l even	$\vec{h}_1$ l_1 odd
$0, 0, 0$	$0, 0, 0$
$0, 0, \frac{1}{2}$	$0, 0, \frac{1}{2}$
$0, \frac{1}{2}, 0$	$0, \frac{1}{2}, 0$
$0, \frac{1}{2}, \frac{1}{2}$	$0, \frac{1}{2}, \frac{1}{2}$
$\frac{1}{2}, 0, 0$	$\frac{1}{2}, 0, 0$
$\frac{1}{2}, 0, \frac{1}{2}$	$\frac{1}{2}, 0, \frac{1}{2}$
$\frac{1}{2}, \frac{1}{2}, 0$	$\frac{1}{2}, \frac{1}{2}, 0$
$\frac{1}{2}, \frac{1}{2}, \frac{1}{2}$	$\frac{1}{2}, \frac{1}{2}, \frac{1}{2}$

Table 7. Sets of centers which give rise to the same value of the structure factor for each of the eight possible sets of indices for space groups of Type $3P_2$. Broken lines separate non-equivalent origins.

$\vec{h}$	$\vec{h}_1$	$\vec{h}_2$	$\vec{h}_{12}$	$\vec{h}_3$	$\vec{h}_{13}$	$\vec{h}_{23}$	$\vec{h}_{123}$
h even	h_1 odd	h_2 odd	h_{12} even	h_3 odd	h_{13} even	h_{23} even	h_{123} odd
k even	k_1 odd	k_2 even	k_{12} odd	k_3 even	k_{13} odd	k_{23} even	k_{123} odd
l even	l_1 even	l_2 odd	l_{12} odd	l_3 even	l_{13} even	l_{23} odd	l_{123} odd
$0, 0, 0$	$0, 0, 0$	$0, 0, 0$	$0, 0, 0$	$0, 0, 0$	$0, 0, 0$	$0, 0, 0$	$0, 0, 0$
$\frac{1}{2}, \frac{1}{2}, \frac{1}{2}$	$\frac{1}{2}, \frac{1}{2}, \frac{1}{2}$	$\frac{1}{2}, \frac{1}{2}, \frac{1}{2}$	$\frac{1}{2}, \frac{1}{2}, \frac{1}{2}$	$0, 0, \frac{1}{2}$	$0, 0, \frac{1}{2}$	$\frac{1}{2}, \frac{1}{2}, 0$	$\frac{1}{2}, \frac{1}{2}, 0$
$0, 0, \frac{1}{2}$	$0, 0, \frac{1}{2}$	$0, \frac{1}{2}, 0$	$\frac{1}{2}, 0, 0$	$0, \frac{1}{2}, 0$	$\frac{1}{2}, 0, \frac{1}{2}$	$0, \frac{1}{2}, 0$	$\frac{1}{2}, 0, \frac{1}{2}$
$\frac{1}{2}, \frac{1}{2}, 0$	$\frac{1}{2}, \frac{1}{2}, 0$	$\frac{1}{2}, 0; \frac{1}{2}$	$0, \frac{1}{2}, \frac{1}{2}$	$0, \frac{1}{2}, \frac{1}{2}$	$\frac{1}{2}, 0, 0$	$\frac{1}{2}, 0, 0$	$0, \frac{1}{2}, \frac{1}{2}$
$0, \frac{1}{2}, 0$	$0, \frac{1}{2}, 0$	$0, 0, \frac{1}{2}$	$0, 0, \frac{1}{2}$	$\frac{1}{2}, \frac{1}{2}, \frac{1}{2}$	$\frac{1}{2}, \frac{1}{2}, \frac{1}{2}$	$\frac{1}{2}, \frac{1}{2}, \frac{1}{2}$	$\frac{1}{2}, \frac{1}{2}, \frac{1}{2}$
$\frac{1}{2}, 0, \frac{1}{2}$	$\frac{1}{2}, 0, \frac{1}{2}$	$\frac{1}{2}, \frac{1}{2}, 0$	$\frac{1}{2}, \frac{1}{2}, 0$	$\frac{1}{2}, \frac{1}{2}, 0$	$\frac{1}{2}, \frac{1}{2}, 0$	$0, 0, \frac{1}{2}$	$0, 0, \frac{1}{2}$
$\frac{1}{2}, 0, 0$	$\frac{1}{2}, 0, 0$	$\frac{1}{2}, 0, 0$	$0, \frac{1}{2}, 0$	$\frac{1}{2}, 0, \frac{1}{2}$	$0, \frac{1}{2}, 0$	$\frac{1}{2}, 0, \frac{1}{2}$	$0, \frac{1}{2}, 0$
$0, \frac{1}{2}, \frac{1}{2}$	$0, \frac{1}{2}, \frac{1}{2}$	$0, \frac{1}{2}, \frac{1}{2}$	$\frac{1}{2}, 0, \frac{1}{2}$	$\frac{1}{2}, 0, 0$	$0, \frac{1}{2}, \frac{1}{2}$	$0, \frac{1}{2}, \frac{1}{2}$	$\frac{1}{2}, 0, 0$

Table 8. For each of the four equivalence classes, sets of centers which give rise to the same value of the structure factor for each of the two possible sets of indices for space groups of Type $3P_2$.

$\vec{h}$ $h+k+l$ even	$\vec{h}_1$ $h_1+k_1+l_1$ odd
$0 \quad 0, \quad 0$	$0, \quad 0, \quad 0$
$\frac{1}{2}, \quad \frac{1}{2}, \quad \frac{1}{2}$	$\frac{1}{2}, \quad \frac{1}{2}, \quad \frac{1}{2}$
$0, \quad 0, \quad \frac{1}{2}$	$0, \quad 0, \quad \frac{1}{2}$
$\frac{1}{2}, \quad \frac{1}{2}, \quad 0$	$\frac{1}{2}, \quad \frac{1}{2}, \quad 0$
$0, \quad \frac{1}{2}, \quad 0$	$0, \quad \frac{1}{2}, \quad 0$
$\frac{1}{2}, \quad 0, \quad \frac{1}{2}$	$\frac{1}{2}, \quad 0, \quad \frac{1}{2}$
$\frac{1}{2}, \quad 0, \quad 0$	$\frac{1}{2}, \quad 0, \quad 0$
$0, \quad \frac{1}{2}, \quad \frac{1}{2}$	$0, \quad \frac{1}{2}, \quad \frac{1}{2}$

linearly independent modulo 2, is originally selected.

Type 2P (Category 2). For each space group there are two forms for the structure factor depending on the choice of origin. Hence there are two equivalence classes, shown in row 6 of Table 1. Our previous discussion of Table 2 applies also to Table 3, but the vectors $\vec{h}_1$, $\vec{h}_2$, $\vec{h}_3$, linearly independent modulo 2, are different. The broken lines in the solid boxes of Table 3 separate the non-equivalent origins. Inspection of Table 3 leads to Table 4 where the double line separates the equivalence classes. For each equivalence class and each index heading a given column of Table 4, all those centers referred to which the corresponding structure factor has the same sign are enclosed in the same box. For each fixed form of structure factor, a discussion analogous to that of Table 2 applies to Table 4.

Types $3P_1$ and $3P_2$ (Category 3). For each space group there are four forms for the structure factor depending on the choice of origin. Hence there are four equivalence classes, shown in row 6 of Table 1. In view of the previous discussions for Types 1P and 2P, Tables 5 to 8 are self-explanatory.

This discussion leads to the following definitions:

Definition 1. For each of the four types described in Table 1, the vectors $\vec{h}_i$ and $\vec{h}_s$, associated invariantly and seminvariantly respectively with the phase $\phi_{\vec{h}}$ (or the vector $\vec{h}$), are defined by rows 7 and 8 of Table 1. Note that $(h + k, l)$ is a two-dimensional vector, and (l) and $(h+k+l)$ are one-dimensional vectors.

Definition 2. For each of the four types described in Table 1, a set of phases $\phi_{\vec{h}_j}$ is said to be linearly dependent or independent modulo 2 according as the set of invariantly associated vectors is linearly dependent or independent modulo 2. The phase $\phi_{\vec{h}}$ is linearly dependent modulo 2 on, or linearly independent modulo 2 of, the set of phases $\phi_{\vec{h}_j}$ according as the vector invariantly associated with $\phi_{\vec{h}}$ is linearly dependent modulo 2 on, or linearly independent modulo 2 of, the set of vectors invariantly associated with the set $\phi_{\vec{h}_j}$.

Definition 3. For each of the four types described in Table 1, a set of phases $\phi_{\vec{h}_j}$ is said to be linearly semi-dependent or semi-independent modulo 2 according as the set of seminvariantly associated vectors is linearly dependent or independent modulo 2. The phase $\phi_{\vec{h}}$ is linearly semi-dependent modulo 2 on, or linearly semi-independent modulo 2 of, the set of phases $\phi_{\vec{h}_j}$ according as the vector seminvariantly associated with $\phi_{\vec{h}}$ is linearly dependent modulo 2 on, or linearly independent modulo 2 of, the set of vectors seminvariantly associated with the set $\phi_{\vec{h}_j}$.

Invariants and Seminvariants

Of some importance are the phases which are structure invariants. These are the phases whose values are determined by the structure alone and are independent of the choice of origin. These phases are characterized by vectors $\vec{h}$ which are even. There are eight possible origins associated with three degrees of freedom. Any phase $\phi_{\vec{h}_1}$ which is not a structure invariant has one value when referred to four of these origins and another value when referred to the other four (Tables 2, 3, 5, 7). Hence the value of $\phi_{\vec{h}_1}$ may be specified arbitrarily, thus reducing the number of permissible origins from eight to four, and then any phase which is linearly dependent modulo 2 on $\phi_{\vec{h}_1}$ has the same value for each of these four permissible origins. However any phase $\phi_{\vec{h}_2}$ which is linearly independent modulo 2 of $\phi_{\vec{h}_1}$ may be specified arbitrarily, since it has one value when referred to two of these origins, but a different value when referred to the other two. Once this is done only two of the four possible origins are allowed. Any phase which is linearly dependent modulo 2 on the pair $\phi_{\vec{h}_1}$, $\phi_{\vec{h}_2}$ has the same value for each of these two origins. However, any phase $\phi_{\vec{h}_3}$ which is linearly independent modulo 2 of the pair $\phi_{\vec{h}_1}$, $\phi_{\vec{h}_2}$ has one value when referred to one of these origins and a different value when referred to the other. Hence the value of $\phi_{\vec{h}_3}$ may be arbitrarily chosen and, once this is done, the origin is uniquely specified. All remaining phases, necessarily linearly dependent modulo 2 on the triple $\phi_{\vec{h}_1}$, $\phi_{\vec{h}_2}$, $\phi_{\vec{h}_3}$, are then determined by the structure.

For our purposes, phases which are structure seminvariants are of particular importance. Let us imagine that the form of the structure factor has been fixed. Since, for Type 1P, there is only one form for the structure factor, this is no particular restriction and the preceding paragraph applies also to the structure seminvariants. However, for Type 2P, there are two forms for the structure factor depending on the choice of origin (Table 3). Hence, specifying the form of the structure factor automatically reduces the possible number of origins from eight to four, leaving only two degrees of freedom. Now those phases which are structure seminvariants (not only the structure invariants) are determined by the structure. The structure invariants will have the same value for each of the two possible forms of structure factor, but those structure seminvariants which are not structure invariants have one value for one form of the structure factor and another value for the other form of the structure factor. For each fixed form of structure factor a phase which is not a structure seminvariant has one value when referred to two of the four permissible origins and another value when referred to the other two (Table 4). Hence any such phase $\phi_{\vec{h}_1^1}$ may be specified arbitrarily, thus reducing the number of permissible origins from four to two. Any phase linearly semi-dependent modulo 2 on $\phi_{\vec{h}_1^1}$ has the same value when referred to either of these two permissible origins. However, any phase $\phi_{\vec{h}_2^1}$ linearly semi-independent modulo 2 of $\phi_{\vec{h}_1^1}$ has one value when referred to one of these origins and another when referred to the other. Hence the value of $\phi_{\vec{h}_2^1}$ may be arbitrarily specified, thus fixing the origin uniquely. Then all remaining phases, necessarily linearly semi-dependent modulo 2 on the pair $\phi_{\vec{h}_1^1}$, $\phi_{\vec{h}_2^1}$, are determined by the structure.

A similar discussion applies to Types $3P_1$ and $3P_2$. Since there are now four forms for the structure factor, specifying the form of the structure factor reduces the number of permissible origins from eight to two, with only one degree of freedom (Tables 5 and 7). For each fixed form of structure factor those phases which are structure seminvariants (i.e. linearly semi-dependent modulo 2) are determined by the structure. Those phases which are linearly semi-independent modulo 2, however, have one value when referred to one origin and another when referred to the other (Tables 6 and 8). Any such phase, $\phi_{\vec{h}_1^1}$, may therefore be arbitrarily specified, thus fixing the origin uniquely. Then all remaining phases, necessarily linearly semi-dependent modulo 2 on $\phi_{\vec{h}_1^1}$ are determined.

Preliminary Theorems

The foregoing discussion completes the proof of the following theorems. It may be noted that Theorem 1 includes Theorems 2 to 5. However, the detailed statements help to clarify a somewhat complicated situation.

Theorem 1. The structure invariants are the linear combinations

$$\dot{\sum_j} A_j \phi_{\vec{h}_j} \tag{2.08}$$

where the A_j's are integers satisfying

$$\sum_j A_j \vec{h}_{j_i} \equiv 0 (\mathrm{mod}\ 2), \tag{2.09}$$

$\vec{h}_{j_i}$ is the vector invariantly associated with $\phi_{\vec{h}_j}$, $\phi_{\vec{h}_j}$ is the phase (either 0 or π) of the structure factor $F_{\vec{h}_j}$, and the symbol $\dot{\sum_j}$ means that the sum in (2.08) is to be reduced modulo 2π (so that the value of (2.08) is either 0 or π). The structure seminvariants are the linear combinations

$$\dot{\sum_j} A_j \phi_{\vec{h}_j} \tag{2.10}$$

where the A_j's are integers satisfying

$$\sum_j A_j \vec{h}_{j_s} \equiv 0 (\mathrm{mod}\ 2), \tag{2.11}$$

$\vec{h}_{j_s}$ is the vector seminvariantly associated with $\phi_{\vec{h}_j}$, $\phi_{\vec{h}_j}$

is the phase (either 0 or π) of the structure factor $F_{\vec{h}_j}$, and the symbol $\dot{\sum_j}$ means that the sum in (2.10) is to be reduced modulo 2π (so that the value of (2.10) is either 0 or π).

Theorem 2. In a given crystal the value of any phase which is linearly dependent modulo 2 is determined by the crystal structure (and is thus independent of the choice of the origin, which is of course always assumed to be at a center of symmetry). For a fixed form of structure factor, the value of any phase which is linearly semi-dependent modulo 2 is determined by the crystal structure.

Theorem 3. In a given crystal the value of any phase which is linearly independent modulo 2 may be specified arbitrarily. However, once such a phase, $\phi_{\vec{h}_1}$, has been specified, then any phase, which is linearly dependent modulo 2 on $\phi_{\vec{h}_1}$, is determined by the crystal structure. For a fixed form of structure factor, the value of any phase which is linearly semi-independent modulo 2 may be specified arbitrarily. However, once such a phase, $\phi_{\vec{h}_1^1}$, has been specified, then any phase which is linearly semi-dependent modulo 2 on $\phi_{\vec{h}_1^1}$, is determined by the crystal structure. For crystals of Types $3P_1$ and $3P_2$ every phase is linearly semi-dependent modulo 2 on $\phi_{\vec{h}_1^1}$, and the second halves of the following Theorems 4 and 5 do not apply.

Theorem 4. In a given crystal the values of any two phases which are linearly independent modulo 2 may be specified arbitrarily. However, once such phases, $\phi_{\vec{h}_1}$ and $\phi_{\vec{h}_2}$, have been specified, then any phase, which is linearly dependent modulo 2 on the pair $\phi_{\vec{h}_1}$, $\phi_{\vec{h}_2}$, is determined by the crystal structure. For Types 1P and 2P and for a fixed form of structure factor, the values of any two phases which are linearly semi-independent modulo 2 may be specified arbitrarily. However, once such phases, $\phi_{\vec{h}_1^1}$ and $\phi_{\vec{h}_2^1}$, have been specified, then any phase which is linearly semi-dependent modulo 2 on the pair $\phi_{\vec{h}_1^1}$, $\phi_{\vec{h}_2^1}$, is determined by the crystal structure. For crystals of Type 2P every phase is linearly semi-dependent modulo 2 on the pair $\phi_{\vec{h}_1^1}$, $\phi_{\vec{h}_2^1}$, and the second half of the following Theorem 5 does not apply.

Theorem 5. In a given crystal any three phases which are linearly independent modulo 2 may be specified arbitrarily. However, once such phases have been specified, then any phase is determined by the crystal structure. For crystals of Type 1P, the concepts of invariance and seminvariance coincide.

Theorem 6. For each type, the similarity classes coincide with the equivalence classes.

Chapter 3

PROBABILITIES

Joint Distribution

As has been observed previously (Hauptman and Karle, 1953) the concept of the joint or compound probability distribution appears to be particularly useful in problems involving probabilities of interdependent events; for the probability distribution of a structure factor when certain magnitudes or phases are specified is readily derivable from the joint distribution.* The structure factor for the centrosymmetric crystal is given by (1.01). Denote by $p(\xi_{j1}, \cdots, \xi_{jm})\, d\xi_{j1} \cdots d\xi_{jm}$ the joint probability that $\xi_{j\mu}$ lie in the interval $\xi_{j\mu}$, $\xi_{j\mu} + d\xi_{j\mu}$, for $\mu = 1, 2, \cdots, m$ where

$$\xi_{j\mu} = \xi(x_j, y_j, z_j, \vec{h}_\mu) \tag{3.01}$$

and m is any positive integer. Let $P_1(A_1, \cdots, A_m)\, dA_1 \cdots dA_m$ be the joint probability that $F_{\vec{h}_\mu}$ lie in the interval A_μ, $A_\mu + dA_\mu$, $\mu = 1, 2, \cdots, m$. We prove next the fundamental result (cf. (1.14) and (1.17))

$$P_1(A_1, \cdots, A_m) = \frac{1}{(2\pi)^m} \int_{-\infty}^{\infty} \cdots \int_{-\infty}^{\infty} \exp\left(-i \sum_{\mu=1}^{m} A_\mu w_\mu\right) \times$$

$$\prod_{j=1}^{N/n} q(f_{j1} w_1, \cdots, f_{jm} w_m) dw_1 \cdots dw_m, \tag{3.02}$$

where

*The restriction to primitive unit cells made in the previous sections is now dropped; and the probability theory to be developed is valid regardless of the choice of unit cell.

$$q(f_{j1}w_1, \cdots, f_{jm}w_m) = \int_{-\infty}^{\infty} \cdots \int_{-\infty}^{\infty} p(\xi_{j1}, \cdots, \xi_{jm}) \times$$

$$\exp\left(i \sum_{\mu=1}^{m} f_{j\mu}\xi_{j\mu}w_\mu\right) d\xi_{j1} \cdots d\xi_{jm}, \qquad (3.03)$$

and

$$f_{j\mu} = f_j(h_\mu, k_\mu, l_\mu) = f_j(\vec{h}_\mu). \qquad (3.04)$$

The probability, $Q(A_1, \cdots, A_m)$, that $F_{\vec{h}_\mu}$ be less than A_μ for every $\mu = 1, 2, \cdots, m$ is

$$Q(A_1, \cdots, A_m) = \int_{-\infty}^{\infty} \cdots \int_{-\infty}^{\infty} \prod_{j=1}^{N/n} \left(p(\xi_{j1}, \cdots, \xi_{jm}) d\xi_{j1} \cdots d\xi_{jm}\right) \times$$

$$\prod_{\mu=1}^{m} T(\xi_{1\mu}, \cdots, \xi_{N/n\,\mu}), \qquad (3.05)$$

where

$$T(\xi_{1\mu}, \cdots, \xi_{N/n\,\mu}) = \frac{1}{2} - \frac{1}{2\pi}\int_{-\infty}^{\infty} \frac{\exp[i(F_{\vec{h}_\mu} - A_\mu)w_\mu]\, dw_\mu}{i\, w_\mu}$$

$$= 1 \quad \text{if} \quad F_{\vec{h}_\mu} < A_\mu, \qquad (3.06)$$

$$= 0 \quad \text{if} \quad F_{\vec{h}_\mu} > A_\mu.$$

By differentiating (3.05) successively with respect to $A_1, \cdots, A_m$, we obtain (3.02) and (3.03) since

$$P_1(A_1, \cdots, A_m) = \frac{\partial^m Q(A_1, \cdots, A_m)}{\partial A_1 \cdots \partial A_m}. \qquad (3.07)$$

Probability Distributions for F

Equations (3.02) and (3.03) are the starting point from which the probability distributions for the structure factors may be derived on the basis that certain sets of magnitudes or phases are known. As in the derivation of (3.02) the atoms in the asymmetric unit are assumed to range at random throughout the unit cell except in so far as they are restricted by a knowledge of the magnitudes or the phases of a specified set of structure factors. The formulas to be derived are of two types, those requiring a knowledge of intensities only, and others requiring a knowledge of the phases also.

In order to express (3.02) in a more useful form we first find the Maclaurin expansion of the exponential in (3.03):

$$q(f_{j1}w_1, \cdots, f_{jm}w_m) = \int_{-\infty}^{\infty}\cdots\int_{-\infty}^{\infty} p(\xi_{j1}, \cdots, \xi_{jm}) d\xi_{j1} \cdots d\xi_{jm} \Big\{1 + i\sum_{\mu=1}^{m} f_{j\mu}\xi_{j\mu}w_\mu - \frac{1}{2!}\Big(\sum_{\mu=1}^{m} f_{j\mu}\xi_{j\mu}w_\mu\Big)^2 - \frac{i}{3!}\Big(\sum_{\mu=1}^{m} f_{j\mu}\xi_{j\mu}w_\mu\Big)^3 + \cdots\Big\}. \quad (3.08)$$

The terms of (3.08) are all of the form of a mixed moment

$$m_{\lambda_1 \ldots \lambda_m} = \int_{-\infty}^{\infty}\cdots\int_{-\infty}^{\infty} p(\xi_{j1}, \cdots, \xi_{jm}) \times \xi_{j1}^{\lambda_1} \cdots \xi_{jm}^{\lambda_m} \, d\xi_{j1} \cdots d\xi_{jm}. \quad (3.09)$$

Interpreting (3.09) as an expected value, or average, of $\xi_{j1}^{\lambda_1} \cdots \xi_{jm}^{\lambda_m}$ we infer that

$$m_{\lambda_1 \ldots \lambda_m} = \int_0^1\int_0^1\int_0^1 \xi_{j1}^{\lambda_1}(x,y,z,\vec{h}_1) \cdots \xi_{jm}^{\lambda_m}(x,y,z,\vec{h}_m)dx\,dy\,dz. \quad (3.10)$$

The importance of (3.10) is due to the fact that in evaluating q from (3.08) it is not necessary to have an explicit expression for $p(\xi_{j1}, \cdots, \xi_{jm})$. It is sufficient to evaluate the moments $m_{\lambda_1 \ldots \lambda_m}$ in (3.10), a relatively simple matter once the functions $\xi_{j1}, \cdots, \xi_{jm}$ have been given. It is thus seen that the exact nature of the interdependence of the vector polygons is revealed by the values of the mixed moments (3.10). As a general rule these moments vanish. However, for suitable relationships among the indices h_μ, k_μ, l_μ, $\mu = 1, 2, \cdots, m$, which depend upon the space group, these moments differ from zero. In this way those vector polygons most intimately related to any given one are determined. Our next task is to discover the significant nonvanishing mixed moments (3.10) and to express the probability distributions in terms of them.

We consider first the case that m = 2 and assume that $\vec{h}_1 \neq \vec{h}_2$. Since $m_{01} = m_{10} = m_{11}$* $= m_{03} = m_{30} = 0$, and not both of m_{12}, m_{21} are different from zero so that we may assume $m_{21} = 0$, (3.08) reduces to

$$q(f_{j1}w_1, f_{j2}w_2) = 1 - \frac{1}{2}(f_{j1}^2 m_{20} w_1^2 + f_{j2}^2 m_{02} w_2^2) - \frac{i}{2} f_{j1} f_{j2}^2 m_{12} w_1 w_2^2 + \cdots. \tag{3.11}$$

Thus, using the Maclaurin expansion of the logarithm,

$$\log \prod_{j=1}^{N/n} q(f_{j1}w_1, f_{j2}w_2) = -\frac{1}{2} m_{20} w_1^2 \sum_{j=1}^{N/n} f_{j1}^2 - \frac{1}{2} m_{02} w_2^2 \sum_{j=1}^{N/n} f_{j2}^2 - \frac{i}{2} m_{12} w_1 w_2^2 \sum_{j=1}^{N/n} f_{j1} f_{j2}^2 + \cdots, \tag{3.12}$$

and

$$\prod_{j=1}^{N/n} q(f_{j1}w_1, f_{j2}w_2) = \exp\left(-\frac{1}{2} m_{20} w_1^2 \sum_{j=1}^{N/n} f_{j1}^2 - \frac{1}{2} m_{02} w_2^2 \sum_{j=1}^{N/n} f_{j2}^2\right) \times \left\{1 - \frac{i}{2} m_{12} w_1 w_2^2 \sum_{j=1}^{N/n} f_{j1} f_{j2}^2 + \cdots\right\}. \tag{3.13}$$

*While m_{11} does not necessarily vanish for some of the more complicated space groups, the case $m_{11} \neq 0$ does not appear to lead to significant results and is therefore ignored.

Substituting from (3.13) into (3.02) we obtain

$$P_1(A_1,A_2) = \frac{\exp\left(-\dfrac{A_1^2}{\dfrac{2m_{20}}{n}\sum\limits_1^N f_{j1}^2} - \dfrac{A_2^2}{\dfrac{2m_{02}}{n}\sum\limits_1^N f_{j2}^2}\right)}{2\pi\left(\dfrac{m_{20}}{n}\sum\limits_1^N f_{j1}^2 \cdot \dfrac{m_{02}}{n}\sum\limits_1^N f_{j2}^2\right)^{\frac{1}{2}}} \left\{ 1 \right.$$

$$+ \frac{m_{12}}{2m_{20}^{\frac{1}{2}}\, m_{02}} \cdot \frac{n^{\frac{1}{2}}\sum\limits_1^N f_{j1} f_{j2}^2}{\left(\sum\limits_1^N f_{j1}^2\right)^{\frac{1}{2}} \sum\limits_1^N f_{j2}^2}$$

$$\left. \cdot \frac{A_1}{\dfrac{m_{20}^{\frac{1}{2}}}{n^{\frac{1}{2}}}\left(\sum\limits_1^N f_{j1}^2\right)^{\frac{1}{2}}} \left(\frac{A_2^2}{\dfrac{m_{02}}{n}\sum\limits_1^N f_{j2}^2} - 1\right) + \cdots \right\} \tag{3.14}$$

where j now ranges over all N atoms in the unit cell. For $F_{\vec{h}_2} = A_2$, (3.14) yields (except for a normalizing factor) the probability distribution for $F_{\vec{h}_1}$.

Next we define the normalized structure factor $E_{\vec{h}}$ by means of

$$E_{\vec{h}} = \frac{n^{\frac{1}{2}} F_{\vec{h}}}{m_2^{\frac{1}{2}}\left(\sum\limits_1^N f_j^2\right)^{\frac{1}{2}}} \tag{3.15}$$

so that $E_{\vec{h}}$ and $F_{\vec{h}}$ have the same sign and $E_{\vec{h}}$ may be computed in terms of $F_{\vec{h}}$. Now, rewriting (1.26),

$$\langle F_{\vec{h}}^2 \rangle_{x,y,z} = \frac{m_2}{n} \sum_{j=1}^{N} f_j^2 \tag{3.16}$$

where $\langle F_{\vec{h}}^2 \rangle_{x,y,z} = \langle F_{\vec{h}}^2 \rangle_{\vec{r}}$ is the average value of $F_{\vec{h}}^2$ for fixed $\vec{h}$ as the atoms range randomly and independently throughout the unit cell subject to the conditions imposed by symmetry. Since m_2 is a function of $\vec{h}$, $\langle F_{\vec{h}}^2 \rangle_{\vec{r}}$ also depends on $\vec{h}$. However, it turns out that $\langle F^2_{\vec{h}} \rangle_{\vec{r}}$ takes on only a finite number of different values, e.g., for space group $P2_1/a$, $\langle F^2_{\vec{h}} \rangle_{\vec{r}} = 2 \sum_{j=1}^{N} f_j^2$ if $\vec{h} = (h,0,l)$ or $(0,k,0)$; otherwise $\langle F^2_{\vec{h}} \rangle_{\vec{r}} = \sum_{j=1}^{N} f_j^2$. Two vectors $\vec{h}_1$ and $\vec{h}_2$ will be said to be similar if $\langle F^2_{\vec{h}_1} \rangle_{\vec{r}} = \langle F^2_{\vec{h}_2} \rangle_{\vec{r}}$. Then the totality of vectors $\vec{h}$ may be divided into a finite number of classes such that any two vectors in any class are similar and no vector in any class is similar to any vector in any other class. It is a fundamental result (Preface, Appendix, and Karle and Hauptman, 1953a) that, in general,

$$\langle F^2 \rangle_{\vec{h}_j} = \langle F^2_{\vec{h}} \rangle_{\vec{r}} \tag{3.17}$$

where $\langle F^2 \rangle_{\vec{h}_j}$ is the average value of $F^2_{\vec{h}_j}$ as the $\vec{h}_j$ range uniformly over the vectors in any class, and $\vec{h}$ is any vector in this class. In short, the average of F^2 over h,k,l is the same as the average of F^2 over x,y,z. In view of (3.15) to (3.17)

$$\langle E^2 \rangle_{\vec{h}} = 1, \tag{3.18}$$

where $\vec{h}$ ranges uniformly over the members of any class, or even over all the vectors in reciprocal space. The importance of (3.18) is that it is the basis, in an obvious way, of a procedure for correcting observed intensities for vibrational motion and for putting them on an absolute scale.

The normalized structure factor E is not to be confused with the unitary structure factor U. The probability distribution for E is always approximately $\frac{1}{\sqrt{2\pi}} \exp(-\frac{1}{2}E^2)$ (1.29) regardless of the crystal symmetry or the crystal specimen. The probability distribution for U, on the other hand, depends upon the particular crystal. The maximum value of U is unity and the average value of U^2 tends to zero as the number of atoms in the unit cell approaches infinity. In contrast, the average value of E^2 is always unity and its maximum value is usually several times larger.

The probability distributions are considerably simplified

when referred to the normalized structure factors rather than the structure factors themselves, and will be used exclusively from now on. Eq. (3.14) then becomes

$$P(E_1,E_2) = \frac{\exp\left(-\frac{1}{2}E_1^2 - \frac{1}{2}E_2^2\right)}{2\pi} \times$$

$$\left\{1 + \frac{m_{12}}{2m_{20}^{\frac{1}{2}}\, m_{02}} \cdot \frac{n^{\frac{1}{2}} \sum_1^N f_{j_1} f_{j_2}^2}{\left(\sum_1^N f_{j_1}^2\right)^{\frac{1}{2}} \sum_1^N f_{j_2}^2} E_1(E_2^2 - 1)\right\}, \qquad (3.19)$$

where $P(E_1, E_2)dE_1dE_2$ is the probability that both $E_{\vec{h}_1}$ lie between E_1 and $E_1 + dE_1$ and $E_{\vec{h}_2}$ lie between E_2 and $E_2 + dE_2$. For $E_{\vec{h}_2} = E_2$, (3.19) yields (except for a constant factor) the probability distribution for $E_{\vec{h}_1}$. However, unless $m_{12} \neq 0$, (3.19) is, to the approximation involved in keeping only two terms, an even function of E_1 and therefore yields no information concerning the sign of $E_{\vec{h}_1}$. It is therefore necessary to find in each space group the relationships between $\vec{h}_1$ and $\vec{h}_2$ for which $m_{12} \neq 0$. This routine problem is readily solved for each space group, and the final result depends upon the form of the structure factor. Illustrations will be given later and the relation to the structure invariants and seminvariants will be clarified. Then, except for a normalizing factor (Uspensky, p. 31), (3.19) is the probability distribution for $E_{\vec{h}_1}$ after it is known that $E_{\vec{h}_2}^2$ is equal to E_2^2. It is to be emphasized that, since (3.19) is <u>not</u> an even function of E_1, information concerning the sign of $E_{\vec{h}_1}$ is now available. In fact the probability that $E_{\vec{h}_1}$ be positive, once the values of $|E_{\vec{h}_1}|$ and $|E_{\vec{h}_2}|$ are known, is readily derivable from (3.19) and will be obtained later.

Eq. (3.19) is the first of a long series of similar expressions obtained from (3.02) by letting $m = 2, 3, 4, \cdots$, assuming suitable linear relationships among the vectors $\vec{h}_1, \vec{h}_2, \cdots, \vec{h}_m$ (so that the associated mixed moments do not vanish), and taking as many terms in the Maclaurin expansion of the exponential in (3.03) as are needed to obtain significant results. Since the procedure is the same as in the derivation of (3.19) the most useful of these formulas are listed without further proof.

$$P(E_1,E_2,E_3) = \frac{\exp\left(-\frac{1}{2}E_1^2 - \frac{1}{2}E_2^2 - \frac{1}{2}E_3^2\right)}{(2\pi)^{\frac{3}{2}}}\left\{1 + \frac{m_{111}}{m_{200}^{\frac{1}{2}}m_{020}^{\frac{1}{2}}m_{002}^{\frac{1}{2}}} \cdot \frac{n^{\frac{1}{2}}\sum_1^N f_{j_1}f_{j_2}f_{j_3}}{\left(\sum_1^N f_{j_1}^2\right)^{\frac{1}{2}}\left(\sum_1^N f_{j_2}^2\right)^{\frac{1}{2}}\left(\sum_1^N f_{j_3}^2\right)^{\frac{1}{2}}} E_1E_2E_3\right\}. \quad (3.20)$$

$$P(E_1,E_2,E_3) = \frac{\exp\left(-\frac{1}{2}E_1^2 - \frac{1}{2}E_2^2 - \frac{1}{2}E_3^2\right)}{(2\pi)^{\frac{3}{2}}}\left\{1 + p_1(E_1^2,E_2^2,E_3^2) + \frac{m_{112}}{2m_{200}^{\frac{1}{2}}m_{020}^{\frac{1}{2}}m_{002}} \cdot \frac{n\sum_1^N f_{j_1}f_{j_2}f_{j_3}^2}{\left(\sum_1^N f_{j_1}^2\right)^{\frac{1}{2}}\left(\sum_1^N f_{j_2}^2\right)^{\frac{1}{2}}\left(\sum_1^N f_{j_3}^2\right)} E_1E_2(E_3^2-1)\right\}. \quad (3.21)$$

$$P(E_1,E_2,E_3) = \frac{\exp\left(-\frac{1}{2}E_1^2 - \frac{1}{2}E_2^2 - \frac{1}{2}E_3^2\right)}{(2\pi)^{\frac{3}{2}}}\left\{1 + p_2(E_1^2,E_2^2,E_3^2) + \frac{m_{122}}{4m_{200}^{\frac{1}{2}}m_{020}m_{002}} \cdot \frac{n^{\frac{3}{2}}\sum_1^N f_{j_1}f_{j_2}^2f_{j_3}^2}{\left(\sum_1^N f_{j_1}^2\right)^{\frac{1}{2}}\left(\sum_1^N f_{j_2}^2\right)\left(\sum_1^N f_{j_3}^2\right)} E_1(E_2^2-1)(E_3^2-1)\right\}.$$

(3.22)

$$P(E_1,E_2,E_3,E_4) = \frac{\exp\left(-\frac{1}{2}E_1^2 - \frac{1}{2}E_2^2 - \frac{1}{2}E_3^2 - \frac{1}{2}E_4^2\right)}{4\pi^2} \times$$

$$\left\{1 + p_3(E_1^2,E_2^2,E_3^2,E_4^2) + \frac{m_{1111}}{m_{2000}^{\frac{1}{2}}\, m_{0200}^{\frac{1}{2}}\, m_{0020}^{\frac{1}{2}}\, m_{0002}^{\frac{1}{2}}} \cdot \right.$$

$$\left. \frac{n \sum_1^N f_{j_1}f_{j_2}f_{j_3}f_{j_4}}{\left(\sum_1^N f_{j_1}^2\right)^{\frac{1}{2}}\left(\sum_1^N f_{j_2}^2\right)^{\frac{1}{2}}\left(\sum_1^N f_{j_3}^2\right)^{\frac{1}{2}}\left(\sum_1^N f_{j_4}^2\right)^{\frac{1}{2}}}\, E_1E_2E_3E_4 \right\}. \quad (3.23)$$

$$P(E_1,E_2,E_3,E_4) = \frac{\exp\left(-\frac{1}{2}E_1^2 - \frac{1}{2}E_2^2 - \frac{1}{2}E_3^2 - \frac{1}{2}E_4^2\right)}{4\pi^2} \times$$

$$\left\{1 + p_4(E_1^2,E_2^2,E_3^2,E_4^2) + \frac{m_{1112}}{2m_{2000}^{\frac{1}{2}}\, m_{0200}^{\frac{1}{2}}\, m_{0020}^{\frac{1}{2}}\, m_{0002}} \cdot \right.$$

$$\left. \frac{n^{\frac{3}{2}}\left(\sum_1^N f_{j_1}f_{j_2}f_{j_3}f_{j_4}^2\right) E_1E_2E_3(E_4^2 - 1)}{\left(\sum_1^N f_{j_1}^2\right)^{\frac{1}{2}}\left(\sum_1^N f_{j_2}^2\right)^{\frac{1}{2}}\left(\sum_1^N f_{j_3}^2\right)^{\frac{1}{2}}\left(\sum_1^N f_{j_4}^2\right)} \right\}. \quad (3.24)$$

$$P(E_1,E_2,E_3,E_4) = \frac{\exp\left(-\frac{1}{2}E_1^2 - \frac{1}{2}E_2^2 - \frac{1}{2}E_3^2 - \frac{1}{2}E_4^2\right)}{4\pi^2} \times$$

$$\left\{1 + p_5(E_1^2,E_2^2,E_3^2,E_4^2) + \frac{m_{1122}}{4m_{2000}^{\frac{1}{2}}\, m_{0200}^{\frac{1}{2}}\, m_{0020}\, m_{0002}} \cdot \right.$$

$$\left. \frac{n^2\left(\sum_1^N f_{j_1}f_{j_2}f_{j_3}^2f_{j_4}^2\right) E_1E_2(E_3^2 - 1)(E_4^2 - 1)}{\left(\sum_1^N f_{j_1}^2\right)^{\frac{1}{2}}\left(\sum_1^N f_{j_2}^2\right)^{\frac{1}{2}}\left(\sum_1^N f_{j_3}^2\right)\left(\sum_1^N f_{j_4}^2\right)} \right\}. \quad (3.25)$$

In these equations j ranges from 1 to N and the p's are symmetric polynomials in E_1^2, E_2^2, $\cdots$ the exact forms of which are unimportant since they do not appear in the final formulas for phase determination to be derived later. Additional formulas may easily be obtained by taking higher order terms in the series expansion of (3.03) or by making use of mixed moments other than those listed. However these appear to be less useful than (3.19) to (3.25).

Probabilities for the Sign of F

From formulas (3.19) to (3.25), the joint probability distributions for certain sets of normalized structure factors, the probability that the sign of a structure factor be plus, on the basis that certain magnitudes or phases are known, may be inferred. Denote by $P_+(F)$ the probability that the sign of F be plus and by $P_-(F)$ the probability that the sign of F be minus. Then from (3.19), for example, replacing $\vec{h}_1$ and $\vec{h}_2$ by $\vec{h}$ and $\vec{h}_\mu$ respectively,

$$\frac{P_+(F_{\vec{h}})}{P_-(F_{\vec{h}})} = \frac{1 + \dfrac{m_{12}}{2m_{20}^{\frac{1}{2}}\, m_{02}} \cdot \dfrac{n^{\frac{1}{2}} \sum_1^N f_j f_{j\mu}^2}{\left(\sum_1^N f_j^2\right)^{\frac{1}{2}} \left(\sum_1^N f_{j\mu}^2\right)} |E| (E_\mu^2 - 1)}{1 - \dfrac{m_{12}}{2m_{20}^{\frac{1}{2}}\, m_{02}} \cdot \dfrac{n^{\frac{1}{2}} \sum_1^N f_j f_{j\mu}^2}{\left(\sum_1^N f_j^2\right)^{\frac{1}{2}} \left(\sum_1^N f_{j\mu}^2\right)} |E| (E_\mu^2 - 1)} \tag{3.26}$$

Since

$$P_+(F) + P_-(F) = 1, \tag{3.27}$$

$$P_+(F) = \frac{P_+(F)/P_-(F)}{1 + \dfrac{P_+(F)}{P_-(F)}}, \tag{3.28}$$

and (3.26) implies

$$P_+(F_{\vec{h}}) = \frac{1}{2} + \frac{m_{12}}{4m_{20}^{\frac{1}{2}}\, m_{02}} \cdot \frac{n^{\frac{1}{2}} \sum_{1}^{N} f_j f_{j\mu}^2}{\left(\sum_{1}^{N} f_j^2\right)^{\frac{1}{2}} \left(\sum_{1}^{N} f_{j\mu}^2\right)} |E| (E_\mu^2 - 1). \qquad (3.29)$$

Formula (3.29) gives the probability that $F_{\vec{h}}$ be positive once the magnitudes of $E_{\vec{h}}$ and $E_{\vec{h}_\mu}$ are known to be $|E|$ and $|E_\mu|$ respectively. In a similar way the following formulas are obtained from (3.20) to (3.25), where, as usual, j ranges from 1 to N. From (3.20),

$$P_+(F_{\vec{h}}) = \frac{1}{2} + \frac{m_{111}}{2m_{200}^{\frac{1}{2}}\, m_{020}^{\frac{1}{2}}\, m_{002}^{\frac{1}{2}}} \cdot$$

$$\frac{n^{\frac{1}{2}} (\Sigma f_j f_{j\mu} f_{j\nu}) |E_1|\, E_\mu E_\nu}{(\Sigma f_j^2)^{\frac{1}{2}} (\Sigma f_{j\mu}^2)^{\frac{1}{2}} (\Sigma f_{j\nu}^2)^{\frac{1}{2}}}, \qquad (3.30)$$

where $P_+(F_{\vec{h}})$ is the probability that $F_{\vec{h}}$ be positive once the magnitude of $E_{\vec{h}}$ is known and both the magnitudes and signs of $E_{\vec{h}_\mu}$ and $E_{\vec{h}_\nu}$ are known. While (3.30) is an important formula, its weakness lies in the fact that it requires a previous knowledge of a relatively large number of phases. Therefore it is useless as a starting point in any procedure for phase determination. In fact (3.30), in itself, is insufficient to determine phases since it leads to the solution in which all structure invariants are zero. This emphasizes the need for phase determining formulas (like (3.29)) which do not require a previous knowledge of phases but only of observed intensities. From (3.21),

$$P_+(F_{\vec{h}}) = \frac{1}{2} + \frac{m_{112}}{4m_{200}^{\frac{1}{2}}\, m_{020}^{\frac{1}{2}}\, m_{002}} \cdot$$

$$\frac{n(\Sigma f_j f_{j\mu} f_{j\nu}^2) |E|\, E_\mu (E_\nu^2 - 1)}{(\Sigma f_j^2)^{\frac{1}{2}} (\Sigma f_{j\mu}^2)^{\frac{1}{2}} (\Sigma f_{j\nu}^2)}, \qquad (3.31)$$

where $P_+(F_{\vec{h}})$ is the probability that $F_{\vec{h}}$ be positive once the magnitudes of $E_{\vec{h}}$ and $E_{\vec{h}_\nu}$ are known and both the magnitude and sign of $E_{\vec{h}_\mu}$ are known. Although (3.31) is again useless as a starting point in the procedure for phase determination, it becomes useful as soon as only a few phases have been determined. From (3.22),

$$P_+(F_{\vec{h}}) = \frac{1}{2} + \frac{m_{122}}{8m_{200}^{\frac{1}{2}}\, m_{020}\, m_{002}} \cdot$$

$$\frac{n^{\frac{3}{2}}(\Sigma f_j f_{j\mu}^2 f_{j\nu}^2)\,|E|\,(E_\mu^2 - 1)(E_\nu^2 - 1)}{(\Sigma f_j^2)^{\frac{1}{2}}\,(\Sigma f_{j\mu}^2)\,(\Sigma f_{j\nu}^2)}, \qquad (3.32)$$

where $P_+(F_{\vec{h}})$ is the probability that $F_{\vec{h}}$ be positive once the magnitudes of $E_{\vec{h}}$, $E_{\vec{h}_\mu}$, and $E_{\vec{h}_\nu}$ are known. A special significance is to be attached to (3.32) since it is the probability that a structure factor be positive on the basis that the magnitudes only (and not the signs) of a certain set of structure factors are known. Hence (3.32) and, to a lesser extent, (3.29) form the starting point of the procedure for sign determination to be described. From (3.23),

$$P_+(F_{\vec{h}}) = \frac{1}{2} + \frac{m_{1111}}{2m_{2000}^{\frac{1}{2}}\, m_{0200}^{\frac{1}{2}}\, m_{0020}^{\frac{1}{2}}\, m_{0002}^{\frac{1}{2}}} \cdot$$

$$\frac{n(\Sigma f_j f_{j1} f_{j\mu} f_{j\nu})\,|E|\,E_1 E_\mu E_\nu}{(\Sigma f_j^2)^{\frac{1}{2}}\,(\Sigma f_{j1}^2)^{\frac{1}{2}}\,(\Sigma f_{j\mu}^2)^{\frac{1}{2}}\,(\Sigma f_{j\nu}^2)^{\frac{1}{2}}}, \qquad (3.33)$$

where $P_+(F_{\vec{h}})$ is the probability that $F_{\vec{h}}$ be positive once the magnitude of $E_{\vec{h}}$ is known and both the magnitudes and signs of $E_{\vec{h}_1}$, $E_{\vec{h}_\mu}$, $E_{\vec{h}_\nu}$ are known. From (3.24),

$$P_+(F_{\vec{h}}) = \frac{1}{2} + \frac{m_{1112}}{4m_{2000}^{\frac{1}{2}}\, m_{0200}^{\frac{1}{2}}\, m_{0020}^{\frac{1}{2}}\, m_{0002}} \cdot$$

$$\frac{n^{\frac{3}{2}}(\Sigma f_j f_{j1} f_{j\mu} f_{j\nu}^2)\,|E|\,E_1 E_\mu (E_\nu^2 - 1)}{(\Sigma f_j^2)^{\frac{1}{2}}\,(\Sigma f_{j1}^2)^{\frac{1}{2}}\,(\Sigma f_{j\mu}^2)^{\frac{1}{2}}\,(\Sigma f_{j\nu}^2)}, \qquad (3.34)$$

where $P_+(F_{\vec{h}})$ is the probability that $F_{\vec{h}}$ be positive once the magnitudes of $E_{\vec{h}}$ and $E_{\vec{h}_\nu}$ are known and both the magnitudes and signs of $E_{\vec{h}_1}$ and $E_{\vec{h}_\mu}$ are known. Finally, from (3.25),

$$P_+(F_{\vec{h}}) = \frac{1}{2} + \frac{m_{1122}}{8m_{2000}^{\frac{1}{2}}\, m_{0200}^{\frac{1}{2}}\, m_{0020}\, m_{0002}} \cdot$$

$$\frac{n^2(\Sigma f_j f_{j_1} f_{j\mu}^2 f_{j\nu}^2)|E|\, E_1(E_\mu^2 - 1)(E_\nu^2 - 1)}{(\Sigma f_j^2)^{\frac{1}{2}}\, (\Sigma f_{j_1}^2)^{\frac{1}{2}}\, (\Sigma f_{j\mu}^2)(\Sigma f_{j\nu}^2)}, \qquad (3.35)$$

where $P_+(F_{\vec{h}})$ is the probability that $F_{\vec{h}}$ be positive once the magnitudes of $E_{\vec{h}}$, $E_{\vec{h}_\mu}$, and $E_{\vec{h}_\nu}$ are known and both the magnitude and sign of $E_{\vec{h}_1}$ are known.

Each of the formulas (3.29) to (3.35) is an expression which in general differs only slightly from 1/2. In order to become an effective tool for phase determination these formulas must therefore be modified so as to include many observations. It is plausible to infer from (3.30), for example, that

$$P_+(F_{\vec{h}}) = \frac{1}{2} + \sum_{\mu,\nu} \frac{m_{111}}{2m_{200}^{\frac{1}{2}}\, m_{020}^{\frac{1}{2}}\, m_{002}^{\frac{1}{2}}} \cdot$$

$$\frac{n^{\frac{1}{2}}\, (\Sigma f_j f_{j\mu} f_{j\nu})\, |E_1|\, E_\mu E_\nu}{(\Sigma f_j^2)^{\frac{1}{2}}\, (\Sigma f_{j\mu}^2)^{\frac{1}{2}}\, (\Sigma f_{j\nu}^2)^{\frac{1}{2}}}, \qquad (3.36)$$

where the summation is extended over all indices μ, ν such that $\vec{h}$, $\vec{h}_\mu$, and $\vec{h}_\nu$ are so related, say $\vec{h} = R(\vec{h}_\mu, \vec{h}_\nu)$, that m_{111} does not vanish. Then $P_+(F_{\vec{h}})$ is the probability that $F_{\vec{h}}$ be positive on the basis that the magnitude of $E_{\vec{h}}$ is known and that the magnitudes and signs of all pairs E_{h_μ}, $E_{\vec{h}_\nu}$ are known, where $\vec{h} = R(\vec{h}_\mu, \vec{h}_\nu)$. It is not difficult to prove (3.36) rigorously. The derivation of (3.20) is repeated now in order to find the joint probability distribution of $E_{\vec{h}}$ and all pairs $E_{\vec{h}_\mu}$, $E_{\vec{h}_\nu}$ with $\vec{h} = R(\vec{h}_\mu, \vec{h}_\nu)$. It is readily verified that we obtain a result similar to (3.20) but that the coefficient of E_1 is now replaced by a summation over a large number of similar terms. Eq. (3.36) is thus an immediate consequence of this result. In this connection, however, a slight complication may arise. If a sufficiently large number of vectors $\vec{h}_\mu$, h_ν with $\vec{h} = R(\vec{h}_\mu, \vec{h}_\nu)$ is used, (3.36) could occasionally be negative or greater than unity. This is due to the fact that an insufficient number of terms in the Maclaurin expansion of (3.03) has been used. So far as (3.36) is concerned, the large exponent in (3.03) makes it an unnecessary complication to take additional terms. The only loss in retaining the simplicity of (3.36) is that the exact assertion (3.30) of sign probability is replaced (in the vicinity of $P_+ = 1$ or 0) by the statement that the sign of $F_{\vec{h}}$ is much more likely than not the same as the sign of

$$\sum_{\mu,\nu} \frac{m_{111}}{2m_{200}^{\frac{1}{2}}\, m_{020}^{\frac{1}{2}}\, m_{002}^{\frac{1}{2}}} \cdot \frac{n^{\frac{1}{2}} (\Sigma f_j f_{j\mu} f_{j\nu})\, E_\mu E_\nu}{(\Sigma f_j^2)^{\frac{1}{2}} (\Sigma f_{j\mu}^2)^{\frac{1}{2}} (\Sigma f_{j\nu}^2)^{\frac{1}{2}}}, \tag{3.37}$$

where $\vec{h} = R(\vec{h}_\mu, \vec{h}_\nu)$. Since our primary aim is a procedure for phase determination we shall be chiefly concerned with expression (3.37) and similar ones obtained in the same way from (3.29) and (3.31) to (3.35).

For each space group it is possible to derive in a routine fashion those relationships among the indices which give rise to non-vanishing values of the mixed moments. This leads in a natural way to a specific procedure for phase determination for each space group. The details are illustrated for space groups $P\bar{1}$ and $P2_1/a$.

Chapter 4

PROCEDURE FOR PHASE DETERMINATION

Space Group $P\bar{1}$

For this space group n = 2 and

$$\xi_{j\mu} = 2 \cos 2\pi(h_\mu x_j + k_\mu y_j + l_\mu z_j). \qquad (4.01)$$

The mixed moments appearing in (3.29) to (3.35) are readily evaluated and are different from zero only if the relationships among the indices shown in Table 9 are satisfied.

The following theorems, the probability counterparts of Theorems 1 to 5, are now easily verified for this space group:

Theorem 1A. If a sufficiently large set of intensities is known then the probability that the sign of any structure invariant be positive differs from 1/2.

Theorem 2A. If a sufficiently large set of intensities is known then $P_+(F_{\vec{h}})$ differs from 1/2 provided that the phase $\phi_{\vec{h}}$ of $F_{\vec{h}}$ is linearly dependent modulo 2.

Theorem 3A. If a sufficiently large set of intensities is known and the sign of $F_{\vec{h}_1}$ has been specified arbitrarily, where the phase $\phi_{\vec{h}_1}$ of $F_{\vec{h}_1}$ is any phase which is linearly independent modulo 2, then $P_+(F_{\vec{h}})$ differs from 1/2 provided that $\phi_{\vec{h}}$, the phase of $F_{\vec{h}}$, is linearly dependent modulo 2 on $\phi_{\vec{h}_1}$.

Theorem 4A. If a sufficiently large set of intensities is known and the signs of $F_{\vec{h}_1}$ and $F_{\vec{h}_2}$ have been specified arbitrarily, where the phases $\phi_{\vec{h}_1}$ and $\phi_{\vec{h}_2}$ of $F_{\vec{h}_1}$ and $F_{\vec{h}_2}$ respectively constitute a pair which is linearly independent modulo 2, then $P_+(F_{\vec{h}})$ differs from 1/2 provided that $\phi_{\vec{h}}$, the phase of $F_{\vec{h}}$, is linearly dependent modulo 2 on the pair $\phi_{\vec{h}_1}$, $\phi_{\vec{h}_2}$.

Theorem 5A. If a sufficiently large set of intensities is known and the signs of $F_{\vec{h}_1}$, $F_{\vec{h}_2}$, and $F_{\vec{h}_3}$ have been

Table 9. Values of non-vanishing mixed moments for $P\bar{1}$, and the relationships among the indices which give rise to the moments. The subscript on each mixed moment is an ordered set of integers each of which refers to the corresponding index ordered by means of the relationship listed, e.g., $m_{112} = m_{\lambda\lambda_\mu\lambda_\nu}$ or $m_{\lambda\lambda_1\lambda_\mu}$, $m_{1122} = m_{\lambda\lambda_1\lambda_\mu\lambda_\nu}$, etc.

Mixed Moment	m_2	m_{12}	m_{111}	m_{112}	m_{122}
Relationship		$\vec{h} = \pm 2\vec{h}_\mu$	$\vec{h} = \pm \vec{h}_\mu \pm \vec{h}_\nu$	$\vec{h} = \pm \vec{h}_\mu \pm 2\vec{h}_\nu$ or $\vec{h} \pm \vec{h}_1 = \pm 2\vec{h}_\mu$	$\vec{h} = \pm 2\vec{h}_\mu \pm 2\vec{h}_\nu$
Value	2	2	2	2	2

Mixed Moment	m_{1111}	m_{1112}	m_{1122}
Relationship	$\vec{h} \pm \vec{h}_1 = \pm \vec{h}_\mu \pm \vec{h}_\nu$	$\vec{h} \pm \vec{h}_1 = \vec{h}_\mu \pm 2\vec{h}_\nu$ or $\vec{h} \pm \vec{h}_1 \pm \vec{h}_2 = 2\vec{h}_\mu$	$\vec{h} \pm \vec{h}_1 = \pm 2\vec{h}_\mu \pm 2\vec{h}_\nu$
Value	2	2	2

arbitrarily specified, where the phases $\phi_{\vec{h}_1}$, $\phi_{\vec{h}_2}$, and $\phi_{\vec{h}_3}$ of $F_{\vec{h}_1}$, $F_{\vec{h}_2}$, and $F_{\vec{h}_3}$ respectively constitute a set which is linearly independent modulo 2, then $P_+(F_{\vec{h}})$ differs from 1/2 for any structure factor $F_{\vec{h}}$.

Theorems 1 to 5, 1A to 5A and (3.29) to (3.35) form the basis for a routine procedure for determining the signs of the structure factors. It is assumed that the magnitudes $|F|$ of the structure factors have been adjusted to an absolute scale and for vibrational motion, e.g. by means of well known averaging procedures. Then the "normalized" structure factor magnitudes $|E|$ may be readily computed from (3.15). The sign of any E is seen from (3.15) to be the same as that of its corresponding F. The E's are arranged in decreasing order and their signs (within each step) will be generally determined in this order.

Step 1. In accordance with Theorem 2A we determine the signs of all structure factors whose phases are linearly dependent modulo 2. This is accomplished by using (3.29) to (3.32). The sign of $E_{\vec{h}}$, where $\vec{h}$ is even, is the sign of

$$\Sigma = \Sigma_1 + \Sigma_2 + \Sigma_3 + \Sigma_4, \tag{4.02}$$

where*†

$$\sum_1 = \sum_{\vec{h}=2\vec{h}_\mu} \frac{\sum_j f_{j\vec{h}}\, f^2_{j\vec{h}_\mu}}{4\left(\sum_j f^2_{j\vec{h}}\right)^{\frac{1}{2}}\left(\sum_j f^2_{j\vec{h}_\mu}\right)} \left(E^2_{\vec{h}_\mu} - 1\right), \tag{4.03}$$

$$\sum_2 = \sum_{\vec{h}=\vec{h}_\mu \pm \vec{h}_\nu} \frac{\sum_j f_{j\vec{h}}\, f_{j\vec{h}_\mu}\, f_{j\vec{h}_\nu}}{2\left(\sum_j f^2_{j\vec{h}}\right)^{\frac{1}{2}}\left(\sum_j f^2_{j\vec{h}_\mu}\right)^{\frac{1}{2}}\left(\sum_j f^2_{j\vec{h}_\nu}\right)^{\frac{1}{2}}} E_{h_\mu} E_{h_\nu}, \tag{4.04}$$

*Eq. (4.03) may be compared with the Harker-Kasper (1948) inequality (8).

†Eq. (4.04) should be compared to (1.3) of Sayre (1952), (18) of Cochran (1952), or (8) of Zachariasen (1952) and to the inequality (34) of Karle and Hauptman (1950).

$$\sum_3 = \sum_{\vec{h}=\vec{h}_\mu \pm 2\vec{h}_\nu} \frac{\sum_j f_{j\vec{h}} f_{j\vec{h}_\mu} f^2_{j\vec{h}_\nu}}{4\left(\sum_j f^2_{j\vec{h}}\right)^{\frac{1}{2}}\left(\sum_j f^2_{j\vec{h}_\mu}\right)^{\frac{1}{2}}\left(\sum_j f^2_{j\vec{h}_\nu}\right)} \times$$

$$E_{\vec{h}_\mu}(E^2_{\vec{h}_\nu} - 1), \qquad (4.05)$$

$$\sum_4 = \sum_{\vec{h}=2\vec{h}_\mu \pm 2\vec{h}_\nu} \frac{\sum_j f_{j\vec{h}} f^2_{j\vec{h}_\mu} f^2_{j\vec{h}_\nu}}{8\left(\sum_j f^2_{j\vec{h}}\right)^{\frac{1}{2}}\left(\sum_j f^2_{j\vec{h}_\mu}\right)\left(\sum_j f^2_{j\vec{h}_\nu}\right)} \times$$

$$(E^2_{\vec{h}_\mu} - 1)(E^2_{\vec{h}_\nu} - 1). \qquad (4.06)$$

Since, initially, only the magnitudes of the E's are known, only Σ_1 (which contains only one summand) and Σ_4 can contribute to Σ in (4.02). However, as soon as a few signs become available, Σ_3 begins to play a role and, as more and more signs become known, Σ_2 plays more and more important a role.

Step 2. In accordance with Theorems 3 and 3A, we specify arbitrarily the sign of the largest normalized structure factor $E_{\vec{h}_1}$, whose phase $\phi_{\vec{h}_1}$ is linearly independent modulo 2, and then determine the signs of all structure factors $F_{\vec{h}}$ where $\phi_{\vec{h}}$ is linearly dependent modulo 2 on $\phi_{\vec{h}_1}$. This is accomplished by using the results of Step 1 and Eqs. (3.30), (3.31), (3.33), (3.34), and (3.35). Let $\vec{h}_{\nu_1} \equiv \vec{h}_1$ (mod 2). The sign of $E_{\vec{h}}$, where both $\vec{h} \pm \vec{h}_{\nu_1}$ are even, is the sign of

$$\Sigma' = \Sigma_2 + \Sigma_3' + \Sigma_5 + \Sigma_6 + \Sigma_7 \qquad (4.07)$$

where Σ_2 is given by (4.04) and

$$\sum_3{}' = \sum_{\vec{h} \pm \vec{h}_{\nu_1} = 2\vec{h}_\nu} \frac{\sum_j f_{j\vec{h}} f_{j\vec{h}_{\nu_1}} f^2_{j\vec{h}_\nu}}{4\left(\sum_j f^2_{j\vec{h}}\right)^{\frac{1}{2}}\left(\sum_j f^2_{j\vec{h}_{\nu_1}}\right)^{\frac{1}{2}}\left(\sum_j f^2_{j\vec{h}_\nu}\right)} \times$$

$$E_{\vec{h}_{\nu_1}}(E^2_{\vec{h}_\nu} - 1) \qquad (4.08)$$

$$\sum_5 = \sum_{\vec{h}\pm\vec{h}_{\nu_1}=\vec{h}_\mu\pm\vec{h}_\nu} \frac{\left(\sum_j f_{j\vec{h}} f_{j\vec{h}_{\nu_1}} f_{j\vec{h}_\mu} f_{j\vec{h}_\nu}\right) E_{\vec{h}_{\nu_1}} E_{\vec{h}_\mu} E_{\vec{h}_\nu}}{2\left(\sum_j f^2_{j\vec{h}}\right)^{\frac{1}{2}}\left(\sum_j f^2_{j\vec{h}_{\nu_1}}\right)^{\frac{1}{2}}\left(\sum_j f^2_{j\vec{h}_\mu}\right)^{\frac{1}{2}}\left(\sum_j f^2_{j\vec{h}_\nu}\right)^{\frac{1}{2}}} \qquad (4.09)$$

$$\sum_6 = \sum_{\vec{h}\pm\vec{h}_{\nu_1}=\vec{h}_\mu\pm 2\vec{h}_\nu} \frac{\sum_j f_{j\vec{h}} f_{j\vec{h}_{\nu_1}} f_{j\vec{h}_\mu} f^2_{j\vec{h}_\nu}}{4\left(\sum_j f^2_{j\vec{h}}\right)^{\frac{1}{2}}\left(\sum_j f^2_{j\vec{h}_{\nu_1}}\right)^{\frac{1}{2}}\left(\sum_j f^2_{j\vec{h}_\mu}\right)^{\frac{1}{2}}\left(\sum_j f^2_{j\vec{h}_\nu}\right)} \times$$

$$E_{\vec{h}_{\nu_1}} E_{\vec{h}_\mu} (E^2_{\vec{h}_\nu} - 1), \qquad (4.10)$$

$$\sum_7 = \sum_{\vec{h}\pm\vec{h}_{\nu_1}=2\vec{h}_\mu\pm 2\vec{h}_\nu} \frac{\sum_j f_{j\vec{h}} f_{j\vec{h}_{\nu_1}} f^2_{j\vec{h}_\mu} f^2_{j\vec{h}_\nu}}{8\left(\sum_j f^2_{j\vec{h}}\right)^{\frac{1}{2}}\left(\sum_j f^2_{j\vec{h}_{\nu_1}}\right)^{\frac{1}{2}}\left(\sum_j f^2_{j\vec{h}_\mu}\right)\left(\sum_j f^2_{j\vec{h}_\nu}\right)} \times$$

$$E_{\vec{h}_{\nu_1}} (E^2_{\vec{h}_\mu} - 1)(E^2_{\vec{h}_\nu} - 1). \qquad (4.11)$$

At first only Σ_3', Σ_5, Σ_6, and Σ_7 are important contributors to Σ' in (4.07) and in computing Σ_5 and Σ_6 use is made of the known signs obtained from Step 1. However, as soon as a few signs are found Σ_5 becomes more important, and as more and more signs become known Σ_2 again plays the dominant role.

Step 3. In accordance with Theorems 4 and 3A we specify arbitrarily the sign of the largest normalized structure factor $E_{\vec{h}_2}$, whose phase $\phi_{\vec{h}_2}$ is linearly independent modulo 2 of $\phi_{\vec{h}_1}$, and then determine the signs of all structure factors $F_{\vec{h}}$ whose phases $\phi_{\vec{h}}$ are linearly dependent modulo 2 on $\phi_{\vec{h}_2}$. This is accomplished as in Step 2, but $\vec{h}_2$ replaces $\vec{h}_1$.

Step 4. In accordance with Theorems 5 and 3A, we specify arbitrarily the sign of the largest normalized structure factor $E_{\vec{h}_3}$, whose phase $\phi_{\vec{h}_3}$ is linearly independent modulo 2 of the pair $\phi_{\vec{h}_1}$, $\phi_{\vec{h}_2}$, and then determine the signs of all structure factors $F_{\vec{h}}$ whose phases are linearly dependent modulo 2 on $\phi_{\vec{h}_3}$. This is accomplished as in Step 2, but $\vec{h}_3$ replaces $\vec{h}_1$.

Step 5. In accordance with Theorem 4A, we determine the signs of all structure factors $F_{\vec{h}}$ whose phases are linearly dependent modulo 2 on the pair $\phi_{\vec{h}_1}, \phi_{\vec{h}_2}$. This is accomplished by using the results of Steps 1, 2, 3, Eqs. (3.30) and (3.33). Let $\vec{h}_{\nu_1} \equiv \vec{h}_1$, $\vec{h}_{\nu_2} \equiv \vec{h}_2$ (mod 2). The sign of $F_{\vec{h}}$, where $\vec{h} \pm \vec{h}_{\nu_1} \pm \vec{h}_{\nu_2}$ are all even, is the sign of

$$\Sigma'' = \Sigma_2 + \Sigma_5' \tag{4.12}$$

where

$$\Sigma_5' = \sum_{\vec{h} \pm \vec{h}_{\nu_1} \pm \vec{h}_{\nu_2} = \vec{h}_\nu} \frac{\left(\sum_j f_{j\vec{h}} f_{j\vec{h}_{\nu_1}} f_{j\vec{h}_{\nu_2}} f_{j\vec{h}_\nu}\right) E_{\vec{h}_{\nu_1}} E_{\vec{h}_{\nu_2}} E_{\vec{h}_\nu}}{2\left(\sum_j f_{j\vec{h}}^2\right)^{\frac{1}{2}} \left(\sum_j f_{j\vec{h}_{\nu_1}}^2\right)^{\frac{1}{2}} \left(\sum_j f_{j\vec{h}_{\nu_2}}^2\right)^{\frac{1}{2}} \left(\sum_j f_{j\vec{h}_\nu}^2\right)^{\frac{1}{2}}}. \tag{4.13}$$

At first the results of Steps 2 and 3 are mainly used, but as more signs are obtained, the results of Step 1 are used more and more.

Step 6. In accordance with Theorem 4A, we determine the signs of all structure factors $F_{\vec{h}}$ whose phases are linearly dependent modulo 2 on the pair $\phi_{\vec{h}_1}, \phi_{\vec{h}_3}$. This is done as in Step 5 using the results of Steps 1, 2, 4, Eqs. (3.30) and (3.33).

Step 7. In accordance with Theorem 4A, we determine the signs of all structure factors $F_{\vec{h}}$ whose phases are linearly dependent modulo 2 on the pair $\phi_{\vec{h}_2}, \phi_{\vec{h}_3}$. This is done as in Step 5 using the results of Steps 1, 3, 4, Eqs. (3.30) and (3.33).

Step 8. In accordance with Theorem 5A, the signs of all remaining structure factors $F_{\vec{h}}$ are determined. This is accomplished by using the results of Steps 1 to 7 and Eq. (3.30). The sign of $F_{\vec{h}}$ is the sign of Σ_2.

Since (3.29) to (3.35) are probability statements the procedure for sign determination based upon them may not always lead to a correct assignment of phase. When necessary, repeated application of the procedure, making use of previously determined phases, may be expected to improve these assignments.

Space Group $P2_1/a$

For this space group n = 4 and

$$\xi_{j\mu} = 4 \cos 2\pi\left(h_\mu x_j + l_\mu z_j + \frac{h_\mu + k_\mu}{4}\right) \times$$

$$\cos 2\pi\left(k_\mu y_j - \frac{h_\mu + k_\mu}{4}\right). \qquad (4.14)$$

The mixed moments appearing in (3.29) to (3.35) are readily evaluated and some of the non-vanishing coefficients are listed in Tables 10 to 12 together with the relationships among the indices which give rise to them. It is evident that Theorems 1A to 5A are valid also for $P2_1/a$. A procedure for phase determination for this space group closely parallels that already described for $P\bar{1}$. The major difference is that the number of summands in each of (4.03) to (4.13) is now greatly increased due to the increased number of relationships among the indices giving rise to non-vanishing values of the mixed moments, as indicated in Tables 10 to 12. This is a consequence of the additional symmetry and results in a corresponding increase in the statistical significance of the various terms. For example, if $\vec{h} = (h, 0, l)$ is linearly dependent modulo 2, then Σ_1 of (4.03) is now replaced by

$$\Sigma_1 = \sum_\mu \frac{\sum_j f^2_{j\vec{h}} f^2_{j\vec{h}_\mu} (-1)^{h_\mu + k_\mu}}{2\sqrt{2}\left(\sum_j f^2_{j\vec{h}}\right)^{\frac{1}{2}}\left(\sum_j f^2_{j\vec{h}_\mu}\right)} (E^2_{h_\mu k_\mu l_\mu} - 1), \qquad (4.15)$$

where $h_\mu = (1/2)h$, $l_\mu = (1/2)l$, provided that k_μ takes on the value zero only if h_μ is even. It is seen that Σ_1 for $P2_1/a$ (Eq. (4.15)) may contain several summands whereas Σ_1 for $P\bar{1}$ (Eq. (4.03)) consists of only a single term.

Table 10. Values of the non-vanishing ratio $m_{12}n^{\frac{1}{2}}/(4m_{20}^{\frac{1}{2}}\,m_{02})$ for $P2_1/a$, and the relationships among the indices which give rise to these values. Case 1 does not include case 2.

Case	Relationships	$\dfrac{m_{12}n^{\frac{1}{2}}}{4m_{20}^{\frac{1}{2}}\,m_{02}}$*
1	$k = 0, \quad h = 2h_\mu, \quad l = 2l_\mu$ or $k = 2k_\mu, \quad h = 0, \quad l = 0$	$\dfrac{(-1)^{h_\mu + k_\mu}}{2\sqrt{2}}$
2	$k = 2k_\mu, \quad h = 2h_\mu, \quad l = 2l_\mu$	$\frac{1}{4}$

*$m_{12} = m_{\lambda\lambda_\mu}$.

Table 11. Values of the non-vanishing ratio $m_{112}n/(4m_{200}^{\frac{1}{2}}\, m_{020}^{\frac{1}{2}}\, m_{002})$ for $P2_1/a$, and the relationships among the indices which give rise to these values. No case is to be included in a preceding case.

Case	Relationships [†‡]	$h + k$ / $h_\mu + k_\mu$ / $h_\nu + k_\nu$ → $\frac{m_{112}\, n^*}{4m_{200}^{\frac{1}{2}}\, m_{020}^{\frac{1}{2}}\, m_{002}}$: even / even / even	even / even / odd	odd / odd / even	odd / odd / odd
1	$k = k_\mu = k_\nu = 0,\ h \pm h_\mu = 2h_\nu,\ l \pm l_\mu = 2l_\nu$	$+\frac{1}{2}$			
2	$k = k_\mu = 0,\ h \pm h_\mu = 2h_\nu,\ l \pm l_\mu = 2l_\nu$	$+\frac{1}{2}$	$-\frac{1}{2}$		
3	$k \pm k_\mu = 0,\ h = l = 0,\ h_\mu = \pm 2h_\nu,\ l_\mu = \pm 2l_\nu$	$+\frac{1}{2}\sqrt{2}$	$-\frac{1}{2}\sqrt{2}$		
4	$k \pm k_\mu = 0,\ h_\mu = l_\nu = 0,\ h = \pm 2h_\nu,\ l = \pm 2l_\nu$	$+\frac{1}{2}\sqrt{2}$	$-\frac{1}{2}\sqrt{2}$		
5	$k \pm k_\mu = 0,\ h \pm h_\mu = 2h_\nu,\ l \pm l_\mu = 2l_\nu$	$+\frac{1}{2}$	$-\frac{1}{2}$	$\pm\frac{1}{2}$ [#]	$\mp\frac{1}{2}$ [#]
6	$k = 0,\ k_\mu = \pm 2k_\nu,\ h \pm h_\mu = 2h_\nu,\ l \pm l_\mu = 2l_\nu$	$+\frac{1}{4}\sqrt{2}$	$+\frac{1}{4}\sqrt{2}$		
7	$k_\mu = 0,\ k = \pm 2k_\nu,\ h \pm h_\mu = 2h_\nu,\ l \pm l_\mu = 2l_\nu$	$+\frac{1}{4}\sqrt{2}$	$+\frac{1}{4}\sqrt{2}$		
8	$k \pm k_\mu = 2k_\nu,\ h \pm h_\mu = 2h_\nu,\ l \pm l_\mu = 2l_\nu$	$+\frac{1}{4}$	$+\frac{1}{4}$	$\pm\frac{1}{4}$ [#]	$\pm\frac{1}{4}$ [#]

*As in Table 9, $m_{112} = m_{\lambda\lambda\mu\lambda\nu}$.

†The role of the k's may be interchanged with that of the h's and l's for each relation, e.g., Case 1 includes also the relation $h = h_\mu = h_\nu = 0$, $l = l_\mu = l_\nu = 0$, $k \pm k_\mu = 2k_\nu$.

‡The upper (lower) signs on h and l relations go together, while those on k relations may be chosen independently of those on h and l relations.

#The upper sign is used when the upper (lower) sign on the corresponding k relation goes with the upper (lower) signs on the h and l relations. The lower sign is used when the upper (lower) sign on the corresponding k relation goes

Table 12. Values of the non-vanishing ratio $m_{122}n^{\frac{3}{2}}/(8m_{200}^{\frac{1}{2}}\, m_{020}\, m_{002})$ for $P2_1/a$, and the relationships among the indices which give rise to these values. No case is to be included in a preceding case.

Case	Relationships[†‡#]	$\frac{m_{122}\, n^{\frac{3}{2}*}}{8m_{200}^{\frac{1}{2}}\, m_{020}\, m_{002}}$			
	$h + k$	even	even	even	even
	$h_\mu + k_\mu$	even	odd	even	odd
	$h_\nu + k_\nu$	even	even	odd	odd
1	$k = 0,\ k_\mu = 0$ or $h_\mu = l_\mu = 0,\ k_\nu = 0,\ h = \pm 2h_\nu,\ l = \pm 2l_\nu$	$+\frac{1}{2}\sqrt{2}$			
2	$k = 0,\ k_\mu = 0$ or $h_\mu = l_\mu = 0,\ h = \pm 2h_\nu,\ l = \pm 2l_\nu$	$+\frac{1}{2}\sqrt{2}$		$-\frac{1}{2}\sqrt{2}$	
3	$k = 0,\ k_\nu = 0,\ h = \pm 2h_\nu,\ l = \pm 2l_\nu$	$+\frac{1}{2}\sqrt{2}$	$+\frac{1}{2}\sqrt{2}$		
4	$k = 0,\ h = \pm 2h_\nu,\ l = \pm 2l_\nu$	$+\frac{1}{2}\sqrt{2}$	$+\frac{1}{2}\sqrt{2}$	$-\frac{1}{2}\sqrt{2}$	$-\frac{1}{2}\sqrt{2}$
5	$k = k_\mu = k_\nu = 0,\ h = 2h_\mu \pm 2h_\nu,\ l = 2l_\mu \pm 2l_\nu$	$+\frac{1}{4}\sqrt{2}$			
6	$k = k_\mu = 0,\ h = 2h_\mu \pm 2h_\nu,\ l = 2l_\mu \pm 2l_\nu$	$+\frac{1}{4}\sqrt{2}$		$-\frac{1}{4}\sqrt{2}$	
**7	$k = 0,\ h = 2h_\mu \pm 2h_\nu,\ l = 2l_\mu \pm 2l_\nu$	$+\frac{1}{4}\sqrt{2}$	$-\frac{1}{4}\sqrt{2}$	$-\frac{1}{4}\sqrt{2}$	$+\frac{1}{4}\sqrt{2}$
8	$k = \pm 2k_\nu,\ k_\mu = 0,\ h = \pm 2h_\mu,\ l = \pm 2l_\mu,\ h_\nu = l_\nu = 0$	$+\frac{1}{2}$			
9	$k = \pm 2k_\nu,\ k_\mu = 0,\ h = \pm 2h_\mu,\ l = \pm 2l_\mu$	$+\frac{1}{2}$		$-\frac{1}{2}$	
10	$k = \pm 2k_\nu,\ k_\mu = 0,\ h = \pm 2h_\nu,\ l = \pm 2l_\nu$	$+\frac{1}{2}$		$+\frac{1}{2}$	
11	$k = \pm 2k_\nu,\ h = \pm 2h_\mu,\ l = \pm 2l_\mu$	$+\frac{1}{2}$	$-\frac{1}{2}$	$-\frac{1}{2}$	$+\frac{1}{2}$
12	$k = \pm 2k_\nu,\ h = \pm 2h_\nu,\ l = \pm 2l_\nu$	$+\frac{1}{2}$	$+\frac{1}{2}$	$+\frac{1}{2}$	$+\frac{1}{2}$
13	$k = \pm 2k_\nu,\ k_\mu = 0,\ h = 2h_\mu \pm 2h_\nu,\ l = 2l_\mu \pm 2l_\nu$	$+\frac{1}{4}$		$+\frac{1}{4}$	
14	$k = \pm 2k_\nu,\ h = 2h_\mu \pm 2h_\nu,\ l = 2l_\mu \pm 2l_\nu$	$+\frac{1}{4}$	$-\frac{1}{4}$	$+\frac{1}{4}$	$-\frac{1}{4}$
15	$k = 2k_\mu \pm 2k_\nu,\ h = 2h_\mu \pm 2h_\nu,\ l = 2l_\mu \pm 2l_\nu$	$+\frac{1}{8}$	$+\frac{1}{8}$	$+\frac{1}{8}$	$+\frac{1}{8}$

*As in Table 11, $m_{122} = m_{\lambda\lambda_\mu\lambda_\nu}$.

†As in Table 11, the role of the k's may be interchanged with that of the h's and l's for each relation.

‡The upper (lower) signs on h and l relations go together, while those on k relations may be chosen independently of those on h and l relations.

#Since the second and third subscripts of m_{122} are identical, μ and ν may be interchanged in each relationship in which μ and ν do not occur symmetrically, e.g., Case 8 includes also the relations $k = \pm 2k_\mu$, $k_\nu = 0$, $h = \pm 2h_\nu$, $l = \pm 2l_\nu$, $h_\mu = l_\mu = 0$. However, when doing this, the fourth and fifth columns must be interchanged.

**Erratum: Case 7 is valid only if $k_\mu \neq \pm k_\nu$. If $k_\mu = \pm k_\nu$, the entries should read $+\frac{3}{8}\sqrt{2}$, $-\frac{1}{8}\sqrt{2}$, $+\frac{3}{8}\sqrt{2}$ in order.

Theorems 1A to 5A, the probability analogues of Theorems 1 to 5 for Type 1P, have already been verified for space groups $P\bar{1}$ and $P2_1/a$. A crucial test of the probability theory is to determine whether the probability analogues of Theorems 1 to 5 for Types 2P, $3P_1$, and $3P_2$ are valid. We consider briefly one space group from each of these three types.

Space Group P4/m

This space group belongs to Type 2P. If the origin is at a center (4/m) then

$$\xi_{j\mu} = 4\left[\cos 2\pi(h_\mu x_j + k_\mu y_j) + \cos 2\pi(k_\mu x_j - h_\mu y_j)\right] \times \cos 2\pi l_\mu z_j. \tag{4.16}$$

If the origin is at a center (2/m) then

$$\xi_{j\mu} = 4\left[\cos 2\pi(h_\mu x_j + k_\mu y_j) \pm \cos 2\pi(k_\mu x_j - h_\mu y_j)\right] \times \cos 2\pi l_\mu z_j, \tag{4.17}$$

and the + or - sign is used according as $h_\mu + k_\mu$ is even or odd. Choosing an origin at a center (4/m) we find

$$m_{12} = 64 \int_0^1\int_0^1\int_0^1 \left[\cos 2\pi(hx + ky) + \cos 2\pi(kx - hy)\right] \times \cos 2\pi lz \left[\cos 2\pi(h_\mu x + k_\mu y) + \cos 2\pi(k_\mu x - h_\mu y)\right]^2 \times \cos^2 2\pi l_\mu z \, dxdydz. \tag{4.18}$$

It is readily verified that m_{12} vanishes unless h + k and l are both even, i.e., unless $\phi_{\bar{h}}$ is linearly semi-dependent modulo 2. Assuming that $\phi_{\bar{h}}$ is linearly semi-dependent modulo 2, i.e., h ± k and l are all even, then m_{12} will not vanish if the following relationship, for example, holds among the indices h,k,l, h_μ, k_μ, l_μ,

$$h_\mu = \frac{1}{2}(h - k), \quad k_\mu = \frac{1}{2}(h + k), \quad l_\mu = \frac{1}{2}l. \tag{4.19}$$

Similar results hold if the origin is chosen at a center (2/m). Detailed examination of the mixed moments m_{111}, m_{112}, m_{122}, etc., is sufficient to verify that the probability analogues of Theorems 1 to 5 are valid also for the space group P4/m of Type 2P.

For space group P4/m, and for all space groups of Type 2P, the procedure for phase determination falls naturally into four steps. The E's are arranged in decreasing order and their signs (within each step) will be generally determined in this order.

Step 1. One of the two possible forms of the structure factor is chosen. In accordance with Theorem 2, we then determine the signs of all structure factors whose phases are linearly semi-dependent modulo 2. This is accomplished by using (3.29) to (3.32). The sign of $E_{\vec{h}}$, where $h \pm k$ and l are all even, is the sign of

$$\Sigma = \Sigma_1 + \Sigma_2 + \Sigma_3 + \Sigma_4, \tag{4.20}$$

where*

$$\Sigma_1 = \sum_{\mu} \frac{m_{12}}{4m_{20}^{\frac{1}{2}} m_{02}} \cdot \frac{n^{\frac{1}{2}} \sum_j f_j f_{j\mu}^2}{\left(\sum_j f_j^2\right)^{\frac{1}{2}} \left(\sum_j f_{j\mu}^2\right)} (E_\mu^2 - 1), \tag{4.21}$$

$$\Sigma_2 = \sum_{\mu,\nu} \frac{m_{111}}{2m_{200}^{\frac{1}{2}} m_{020}^{\frac{1}{2}} m_{002}^{\frac{1}{2}}} \cdot \frac{n^{\frac{1}{2}} \left(\sum_j f_j f_{j\mu} f_{j\nu}\right) E_\mu E_\nu}{\left(\sum_j f_j^2\right)^{\frac{1}{2}} \left(\sum_j f_{j\mu}^2\right)^{\frac{1}{2}} \left(\sum_j f_{j\nu}^2\right)^{\frac{1}{2}}}, \tag{4.22}$$

*For convenience, the subscript $\vec{h}$ is suppressed when there is no danger of ambiguity; e.g., $f_{j\vec{h}}$ is replaced by f_j, $f_{j\vec{h}_\mu}$ by $f_{j\mu}$, $E_{\vec{h}_\mu}$ by E_μ, etc.

$$\sum_3 = \sum_{\mu,\nu} \frac{m_{112}}{4m_{200}^{\frac{1}{2}}\, m_{020}^{\frac{1}{2}}\, m_{002}} \cdot$$

$$\frac{n\left(\sum_j f_j\, f_{j\mu}\, f_{j\nu}^2\right) E_\mu\ (E_\nu^2 - 1)}{\left(\sum_j f_j^2\right)^{\frac{1}{2}} \left(\sum_j f_{j\mu}^2\right)^{\frac{1}{2}} \left(\sum_j f_{j\nu}^2\right)}, \qquad (4.23)$$

$$\sum_4 = \sum_{\mu,\nu} \frac{m_{122}}{8m_{200}^{\frac{1}{2}}\, m_{020}\, m_{002}} \cdot$$

$$\frac{n^{\frac{3}{2}}\left(\sum_j f_j\, f_{j\mu}^2\, f_{j\nu}^2\right) (E^2_\mu - 1)(E_\nu^2 - 1)}{\left(\sum_j f_j^2\right)^{\frac{1}{2}} \left(\sum_j f_{j\mu}^2\right) \left(\sum_j f_{j\nu}^2\right)}, \qquad (4.24)$$

and the summations are extended over those values of the indices corresponding to which the associated mixed moments m_{12}, m_{111}, m_{112}, m_{122} do not vanish. For example, for space group P4/m, (4.21) will contain, among possible others, the term with $h_\mu = \frac{1}{2}(h - k)$, $k_\mu = \frac{1}{2}(h + k)$, $l_\mu = \frac{1}{2}l$, as (4.19) shows. Since, initially, only the magnitudes of the E's are known, only Σ_1 and Σ_4 can contribute to Σ at first. However, as soon as a few signs become available, Σ_3 begins to play a role and, as more and more signs become known, Σ_2 plays more and more important a role.

Step 2. In accordance with Theorem 3, we specify arbitrarily the sign of the largest normalized structure factor $E_{\vec{h}_1^1}$ whose phase $\phi_{\vec{h}_1^1}$ is linearly semi-independent modulo 2, and then determine the signs of all structure factors $F_{\vec{h}}$ where $\phi_{\vec{h}}$ is linearly semi-dependent modulo 2 on $\phi_{\vec{h}_1^1}$. This is accomplished by using the results of Step 1 and (3.30), (3.31), (3.33), (3.34), and (3.35). Let $\phi_{\vec{h}_{\nu_1^1}}$ be linearly semi-dependent modulo 2 on $\phi_{\vec{h}_1^1}$. The sign of $E_{\vec{h}}$, where $(h \pm h_{\nu_1^1}) \pm (k \pm k_{\nu_1^1})$ and $l \pm l_{\nu_1^1}$ are all even, is the sign of

$$\Sigma' = \Sigma_2 + \Sigma_3' + \Sigma_5 + \Sigma_6 + \Sigma_7 \tag{4.25}$$

where Σ_2 is given by (4.22) and

$$\sum_3' = \sum_{\nu_1^1, \nu} \frac{m_{112}}{4 m_{200}^{\frac{1}{2}} m_{020}^{\frac{1}{2}} m_{002}} \cdot \frac{n\left(\sum_j f_j f_{j\nu_1^1} f_{j\nu}^2\right) E_{\nu_1^1} (E_\nu^2 - 1)}{\left(\sum_j f_j^2\right)^{\frac{1}{2}} \left(\sum_j f_{j\nu_1^1}^2\right)^{\frac{1}{2}} \left(\sum_j f_{j\nu}^2\right)}, \tag{4.26}$$

$$\sum_5 = \sum_{\nu_1^1, \mu, \nu} \frac{m_{1111}}{2 m_{2000}^{\frac{1}{2}} m_{0200}^{\frac{1}{2}} m_{0020}^{\frac{1}{2}} m_{0002}^{\frac{1}{2}}} \cdot \frac{n\left(\sum_j f_j f_{j\nu_1^1} f_{j\mu} f_{j\nu}\right) E_{\nu_1^1} E_\mu E_\nu}{\left(\sum_j f_j^2\right)^{\frac{1}{2}} \left(\sum_j f_{j\nu_1^1}^2\right)^{\frac{1}{2}} \left(\sum_j f_{j\mu}^2\right)^{\frac{1}{2}} \left(\sum_j f_{j\nu}^2\right)^{\frac{1}{2}}}, \tag{4.27}$$

$$\sum_6 = \sum_{\nu_1^1, \mu, \nu} \frac{m_{1112}}{4 m_{2000}^{\frac{1}{2}} m_{0200}^{\frac{1}{2}} m_{0020}^{\frac{1}{2}} m_{0002}} \cdot \frac{n^{\frac{3}{2}}\left(\sum_j f_j f_{j\nu_1^1} f_{j\mu} f_{j\nu}^2\right) E_{\nu_1^1} E_\mu (E_\nu^2 - 1)}{\left(\sum_j f_j^2\right)^{\frac{1}{2}} \left(\sum_j f_{j\nu_1^1}^2\right)^{\frac{1}{2}} \left(\sum_j f_{j\mu}^2\right)^{\frac{1}{2}} \left(\sum_j f_{j\nu}^2\right)}, \tag{4.28}$$

$$\Sigma_7 = \sum_{\nu_1^1, \mu, \nu} \frac{m_{1122}}{8 m_{2000}^{\frac{1}{2}} m_{0200}^{\frac{1}{2}} m_{0020} m_{0002}}$$

$$\frac{n^2 \left(\sum_j f_j f_{j\nu_1^1} f_{j\mu}^2 f_{j\nu}^2 \right) E_{\nu_1^1} (E_\mu^2 - 1)(E_\nu^2 - 1)}{\left(\sum_j f_j^2 \right)^{\frac{1}{2}} \left(\sum_j f_{j\nu_1^1}^2 \right)^{\frac{1}{2}} \left(\sum_j f_{j\mu}^2 \right) \left(\sum_j f_{j\nu}^2 \right)} \qquad (4.29)$$

and the summations are extended over those values of the indices corresponding to which the associated mixed moments m_{112}, m_{1111}, m_{1112}, m_{1122} do not vanish. At first only Σ_3', Σ_5, Σ_6, and Σ_7 are important contributors to Σ' in (4.25), and in computing Σ_5 and Σ_6 use is made of the known signs obtained from Step 1. However, as soon as a few signs are found Σ_5 becomes more important, and as more and more signs become known Σ_2 again plays the dominant role.

Step 3. In accordance with Theorems 3 and 4, we specify arbitrarily the sign of the largest normalized structure factor $E_{\vec{h}_2^1}$, whose phase $\phi_{\vec{h}_2^1}$ is linearly semi-independent modulo 2 of $\phi_{\vec{h}_1^1}$, and then determine the signs of all structure factors $F_{\vec{h}}$ whose phases $\phi_{\vec{h}}$ are linearly semi-dependent modulo 2 on $\phi_{\vec{h}_2^1}$. This is accomplished as in Step 2, but $\vec{h}_2^1$ replaces $\vec{h}_1^1$.

Step 4. In accordance with Theorem 4 the signs of all remaining structure factors $F_{\vec{h}}$, whose phases $\phi_{\vec{h}}$ are of necessity linearly semi-dependent modulo 2 on the pair $\phi_{\vec{h}_1^1}$, $\phi_{\vec{h}_2^1}$ are determined. This is accomplished by using the results of Steps 1, 2, 3 and Eqs. (3.30) and (3.33). Let $\phi_{\vec{h}_{\nu_1^1}}$ be linearly semi-dependent modulo 2 on $\phi_{\vec{h}_1^1}$ and let $\phi_{\vec{h}_{\nu_2^1}}$ be linearly semi-dependent modulo 2 on $\phi_{\vec{h}_2^1}$. The sign of $F_{\vec{h}}$, where $(h \pm h_{\nu_1^1} \pm h_{\nu_2^1}) \pm (k \pm k_{\nu_1^1} \pm k_{\nu_2^1})$ and $(l \pm l_{\nu_1^1} \pm l_{\nu_2^1})$ are all even, is the sign of

$$\Sigma'' = \Sigma_2 + \Sigma_5' \qquad (4.30)$$

where

$$\sum_5{}' = \sum_{\nu_1^1, \nu_2^1, \nu} \frac{m_{1111}}{m_{2000}^{\frac{1}{2}} m_{0200}^{\frac{1}{2}} m_{0020}^{\frac{1}{2}} m_{0002}^{\frac{1}{2}}}$$

$$\frac{n\left(\sum_j f_j f_{j\nu_1^1} f_{j\nu_2^1} f_{j\nu}\right) E_{\nu_1^1} E_{\nu_2^1} E_{\nu}}{\left(\sum_j f_j^2\right)^{\frac{1}{2}} \left(\sum_j f_{j\nu_1^1}^2\right)^{\frac{1}{2}} \left(\sum_j f_{j\nu_2^1}^2\right)^{\frac{1}{2}} \left(\sum_j f_{j\nu}^2\right)^{\frac{1}{2}}}, \qquad (4.31)$$

and the summation is extended over those indices corresponding to which the associated mixed moment m_{1111} does not vanish.

Space Group $P\bar{3}$

This space group belongs to Type $3P_1$. Choosing the origin at a center ($\bar{3}$) we have

$$\begin{aligned} \xi_{j\mu} = 2\,[&\cos 2\pi(h_\mu x_j + k_\mu y_j + l_\mu z_j) + \\ &\cos 2\pi(k_\mu x_j + i_\mu y_j + l_\mu z_j) + \\ &\cos 2\pi(i_\mu x_j + h_\mu y_j + l_\mu z_j)] \end{aligned} \qquad (4.32)$$

and

$$\begin{aligned} m_{12} = 8 \int_0^1\int_0^1\int_0^1 &[\cos 2\pi(hx + ky + lz) + \\ &\cos 2\pi(kx + iy + lz) + \cos 2\pi(ix + hy + lz)] \times \\ &[\cos 2\pi(h_\mu x + k_\mu y + l_\mu z) + \cos 2\pi(k_\mu x + i_\mu y + l_\mu z) + \\ &\cos 2\pi(i_\mu x + h_\mu y + l_\mu z)]^2 \, dxdydz. \end{aligned} \qquad (4.33)$$

Now m_{12} vanishes unless l is even, i.e. unless $\phi_{\vec{h}}$ is linearly semi-dependent modulo 2. Assuming that $\phi_{\vec{h}}$ is linearly semi-dependent modulo 2, i.e. that l is even, then m_{12} will not vanish if the following relationship, for example, holds among the indices $h, k, l, h_\mu, k_\mu, l_\mu$,

$$h_\mu = -h, \quad k_\mu = -k, \quad l_\mu = (1/2)l. \qquad (4.34)$$

Again, examination of the various mixed moments is sufficient to verify the probability analogues of Theorems 1 to 5 for the space group $P\bar{3}$ of Type $3P_1$.

For space group $P\bar{3}$, and for all space groups of Type $3P_1$, the procedure for phase determination falls naturally into two steps. The E's are arranged in decreasing order and their signs (within each step) will be generally determined in this order.

Step 1. One of the four possible forms of the structure factor is chosen. In accordance with Theorem 2, we then determine the signs of all structure factors whose phases are linearly semi-dependent modulo 2. This is accomplished by using (3.29) to (3.32). The sign of $E_{\vec{h}}$, where l is even, is the sign of Σ(4.20) where Σ_1, Σ_2, Σ_3, Σ_4 are given by (4.21) to (4.24). For space group $P\bar{3}$, for example, (4.21) will contain, among possible others, the term with $h_\mu = -h$, $k_\mu = -k$, $l_\mu = \frac{1}{2}l$, as (4.34) shows. As before, only Σ_1 and Σ_4 contribute to Σ at first. However as soon as some signs become available, Σ_2 and Σ_3 become more useful.

Step 2. In accordance with Theorem 3, we specify arbitrarily the sign of the largest normalized structure factor $E_{\vec{h}_1^1}$, whose phase $\phi_{\vec{h}_1^1}$ is linearly semi-independent modulo 2 (i.e. l_1^1 is odd), and then determine the signs of all remaining structure factors $F_{\vec{h}}$ (i.e. l is odd), whose phases $\phi_{\vec{h}}$ are of necessity linearly semi-dependent modulo 2 on $\phi_{\vec{h}_1^1}$. This is accomplished by using the results of Step 1 and (3.30), (3.31), (3.33), (3.34), and (3.35). The sign of $E_{\vec{h}}$, where $l \pm l_{\nu_1^1}$ are both even, is the sign of Σ' (4.25) where Σ_3', Σ_5, Σ_6, Σ_7 are given by (4.26) to (4.29).

Space Group $R\bar{3}$

This space group belongs to Type $3P_2$. Choosing the origin at a center ($\bar{3}$) we have

$$\xi_{j\mu} = 2[\cos 2\pi(h_\mu x_j + k_\mu y_j + l_\mu z_j) + \cos 2\pi(k_\mu x_j + l_\mu y_j + h_\mu z_j) + \cos 2\pi(l_\mu x_j + h_\mu y_j + k_\mu z_j)], \quad (4.35)$$

and

$$m_{12} = 8 \int_0^1\int_0^1\int_0^1 [\cos 2\pi(hx + ky + lz) + \cos 2\pi(kx + ly + hz) + \cos 2\pi(lx + hy + kz)] \times [\cos 2\pi(h_\mu x + k_\mu y + l_\mu z) + \cos 2\pi(k_\mu x + l_\mu y + h_\mu z) + \cos 2\pi(l_\mu x + h_\mu y + k_\mu z)]^2 \, dxdydz. \quad (4.36)$$

It is easily shown that m_{12} vanishes unless $h + k + l$ is even, i.e., unless $\phi_{\vec{h}}$ is linearly semi-dependent modulo 2. Assuming that $\phi_{\vec{h}}$ is linearly semi-dependent modulo 2, i.e. that $\pm h \pm k \pm l$ are all even, then m_{12} will not vanish if the following relationship, for example, holds among the indices $h, k, l, h_\mu, k_\mu, l_\mu$,

$$h_\mu = \frac{1}{2}(-h+k+l), \quad k_\mu = \frac{1}{2}(h-k+l), \quad l_\mu = \frac{1}{2}(h+k-l). \quad (4.37)$$

Finally, detailed study of the various mixed moments suffices to establish the probability analogues of Theorems 1 to 5 for the space group $R\bar{3}$ of Type $3P_2$.

For space group $R\bar{3}$, and for all space groups of Type $3P_2$, the procedure for phase determination falls naturally into two steps. The E's are arranged in decreasing order and their signs (within each step) will be generally determined in this order.

Step 1. One of the four possible forms of the structure factor is chosen. In accordance with Theorem 2, we then determine the signs of all structure factors whose phases are linearly semi-dependent modulo 2. The sign of $E_{\vec{h}}$ where, $\pm h \pm k \pm l$ are all even, is the sign of Σ (4.20) where $\Sigma_1, \Sigma_2, \Sigma_3, \Sigma_4$ are given by (4.21) to (4.24). For space group $R\bar{3}$, for example, (4.21) will contain, among possible others, the term with $h_\mu = \frac{1}{2}(-h + k + l)$, $k_\mu = \frac{1}{2}(h - k + l)$, $l_\mu = \frac{1}{2}(h + k - l)$, as (4.37) shows. Again only Σ_1 and Σ_4 contribute to Σ at first, but Σ_2 and Σ_3 become more important as soon as some signs become available.

Step 2. In accordance with Theorem 3, we specify arbitrarily the sign of the largest normalized structure factor $E_{\vec{h}_1^1}$ whose phase $\phi_{\vec{h}_1^1}$ is linearly semi-independent modulo 2 (i.e. $\pm h_1^1 \pm k_1^1 \pm l_1^1$ are all odd), and then determine the signs of all remaining structure factors $F_{\vec{h}}$ (i.e. $\pm h \pm k \pm l$ are all odd), whose phases $\phi_{\vec{h}}$ are of necessity linearly semi-dependent modulo 2 on $\phi_{\vec{h}_1^1}$. The sign of $E_{\vec{h}}$, where $(\pm h \pm k \pm l) \pm (\pm h_{\nu_1^1} \pm k_{\nu_1^1} \pm l_{\nu_1^1})$ are all even, is the sign of Σ' (4.25) where Σ_3', Σ_5, Σ_6, Σ_7 are given by (4.26) to (4.29).

These examples indicate that the probability analogues of Theorems 1 to 5 are true for all centrosymmetric space groups, i.e. the probability theory leads to those conclusions concerning the invariants and seminvariants as would have been anticipated from the preliminary analysis which did not involve probabilities (Theorems 1 to 5). They constitute strong and independent evidence that the probability theory developed here is valid.

Simplified Procedure

Eqs. (3.29) to (3.36) and the procedure for sign determination may be considerably simplified if we use the relationship

$$f_{j\vec{h}} = Z_j f_{\vec{h}} \tag{4.38}$$

where Z_j is the atomic number of the jth atom and $f_{\vec{h}}$ is a function of $\vec{h}$ which is assumed to be the same for all atoms present. Then, referring for example to (3.34), we obtain

$$\frac{\sum_j f_{j\vec{h}} f_{j\vec{h}_1} f_{j\vec{h}_\mu} f^2_{j\vec{h}_\nu}}{\left(\sum_j f^2_{j\vec{h}}\right)^{\frac{1}{2}} \left(\sum_j f^2_{j\vec{h}_1}\right)^{\frac{1}{2}} \left(\sum_j f^2_{j\vec{h}_\mu}\right)^{\frac{1}{2}} \left(\sum_j f^2_{j\vec{h}_\nu}\right)} = \frac{\sum_j Z_j^5}{\left(\sum_j Z_j^2\right)^{\frac{5}{2}}}, \tag{4.39}$$

which is seen to be valid to a good approximation even though the $f_{\vec{h}}$ may differ somewhat for different values of j. This results in a corresponding simplification of (3.34). In a similar manner (3.29) to (3.36) may be simplified thus leading to a considerable reduction in the computations of (4.03) to (4.13), the procedure for sign determination. This is due to the fact that the coefficients involving the atomic structure factors are now replaced by numbers (functions of the atomic numbers) independent of the vectors $\vec{h}$, $\vec{h}_1$, $\vec{h}_\mu$, etc., and may therefore be taken outside the signs of summation.

Example

The data of Abrahams, Robertson, and White (1949) for naphthalene have already been corrected for vibrational motion and placed on an absolute scale (Karle and Hauptman, 1953b). The values of $E^2_{\vec{h}}$ were then computed from (3.15).

The values of E^2_{h0l}, where h and l are both even, are arranged in decreasing order, and the first twenty listed in Table 13. Hence

$$E^2_{h_i0l_i} > E^2_{h_j0l_j}, \quad i = 1,2,\cdots,19, \quad i < j.$$

Using (4.02), (4.03), and (4.05), the functions

$$\sum\nolimits_i = \frac{\sum_{j=1}^{36} Z_j^3}{2\sqrt{2}\left(\sum_{j=1}^{36} Z_j^2\right)^{\frac{3}{2}}} \sum\nolimits_{1i} + \frac{\sum_{j=1}^{36} Z_j^4}{2\left(\sum_{j=1}^{36} Z_j^2\right)^{2}} \sum\nolimits_{3i} \qquad (4.40)$$

$$\sum\nolimits_{1i} = \sum_{\mu} (-1)^{h_\mu + k_\mu} (E^2_{h_\mu k_\mu l_\mu} - 1) \qquad (4.41)$$

Table 13. The signs of twenty E_{h0l}'s computed from (4.40).

$\vec{h}$	E^2	Σ_{1i}	Σ_{3i}	Σ_i^*	Expected signs (Abrahams et al.)
$60\bar{8}$	6.85	+1.4	0.0	+0.08	+
$80\bar{8}$	3.74	+1.0	+15.1	+0.42	+
200	3.25	+2.5	+10.9	+0.40	+
$80\bar{4}$	1.91	+4.6	+28.4	+0.93	+
$60\bar{4}$	1.90	+3.7	+49.6	+1.39	+
$40\bar{8}$	1.65	+1.2	+35.1	+0.91	+
004	1.55	+5.2	+33.7	+1.09	+
$20\overline{10}$	1.11	+2.0	+29.0	+0.80	+
600	1.04	-2.8	- 2.5	-0.21	-
208	0.96	+2.5	-13.7	-0.19	+
$40\overline{10}$	0.81	+3.1	+26.9	+0.82	+
$80\bar{2}$	0.73	-3.0	+ 4.3	-0.06	-
206	0.69	0.0	-30.1	-0.72	-
406	0.57	+0.1	-28.2	-0.67	-
$80\overline{10}$	0.56	-2.2	+ 4.1	-0.02	-
$100\bar{8}$	0.55	+1.3	-12.7	-0.23	-
204	0.55	+3.1	+31.5	+0.93	+
402	0.52	+7.5	+ 3.7	+0.50	+
002	0.52	+1.4	-16.7	-0.32	-
$40\bar{4}$	0.47	-2.2	+42.6	+0.90	+

$^*\Sigma_i = 0.055\,\Sigma_{1i} + 0.024\,\Sigma_{3i}$.

where $h_\mu = 1/2\ h_i$, $l_\mu = 1/2\ l_i$,

$$\Sigma_{3i} = \sum_{\substack{\mu, j \\ j < i}} (-1)^{h_\mu + k_\mu} E_{h_j 0 l_j} (E^2_{h_\mu k_\mu l_\mu} - 1) \qquad (4.42)$$

where $h_\mu = 1/2\ (h_i \pm h_j)$, $l_\mu = 1/2\ (l_i \pm l_j)$ and k_μ takes on the value zero only if h_μ is even, were computed for each value of i from 1 to 20. In accordance with Case 1 of Table 10, and Cases 1 and 2 of Table 11, it is to be expected that the sign of $E_{h_i 0 l_i}$ would be the sign of Σ_i. It should be noted that in this example only terms corresponding to Σ_1 and Σ_3 of (4.02) have been used, whereas in the procedure for sign determination two other terms corresponding to Σ_2 and Σ_4 of (4.02) would also be used. The results are shown in Table 13 together with the signs obtained by Abrahams, Robertson, and White. It is seen that only one sign was obtained incorrectly while two others were indecisive. The indecisive results for $E_{80\bar{2}}$ and $E_{80\overline{10}}$ are, in fact, partly due to the use of the previously determined incorrect sign of E_{208}.

As an illustration of the use of (4.41) and (4.42) we rewrite them explicitly for i = 1, 2. If i = 1, (4.41) becomes, since $(h_1 0 l_1) = (60\bar{8})$,

$$\Sigma_{11} = \sum_{\mu} (-1)^{3 + k_\mu} (E^2_{3 k_\mu \bar{4}} - 1), \qquad k_\mu \neq 0,$$

while (4.42) does not contribute to (4.40) since there are no values of j less than 1. If i = 2, (4.41) becomes, since $(h_2 0 l_2) = (80\bar{8})$,

$$\Sigma_{12} = \sum_{\mu} (-1)^{4 + k_\mu} (E^2_{4 k_\mu \bar{4}} - 1).$$

Since i = 2, j takes on the value 1 in (4.42). From Table 13, $E_{60\bar{8}} = +2.617$ and (4.42) becomes

$$\sum_{32} = 2.617\left\{\sum_{\mu}(-1)^{7+k_\mu}(E^2_{7k_\mu 8} - 1) + \right.$$

$$\left.\sum_{\mu}(-1)^{-1+k_\mu}(E^2_{1k_\mu 0} - 1)\right\}, \quad k_\mu \neq 0.$$

Next, the values of E^2_{hkl}, where h, $k \neq 0$, and l are all even, are arranged in decreasing order and the first ten listed in Table 14. Hence $E^2_{h_ik_il_i} > E^2_{h_jk_jl_j}$, $i = 1,2,\cdots,9$, $i < j$. Using (4.02), (4.03), (4.05) and (4.06), the functions

$$\sum_i' = \frac{\sum_{j=1}^{36} Z_j^3}{4\left(\sum_{j=1}^{36} Z_j^2\right)^{\frac{3}{2}}}\sum_{1i}' + \frac{\sqrt{2}\sum_{j=1}^{36} Z_j^4}{4\left(\sum_{j=1}^{36} Z_j^2\right)^{\frac{4}{2}}}\sum_{3i}' +$$

$$\frac{\sum_{j=1}^{36} Z_j^5}{8\left(\sum_{j=1}^{36} Z_j^2\right)^{\frac{5}{2}}}\sum_{4i}', \tag{4.43}$$

$$\sum_{1i}' = (E^2_{h_\mu k_\mu l_\mu} - 1), \tag{4.44}$$

where $\vec{h}_\mu = 1/2\,\vec{h}_i$,

$$\sum_{3i}' = \sum_{j,\mu} E_{h_j0l_j}(E^2_{h_\mu k_\mu l_\mu} - 1), \tag{4.45}$$

where $k_\mu = 1/2\ k_i$, $h_\mu = 1/2\ (h_i \pm h_j)$, $l_\mu = 1/2\ (l_i \pm l_j)$ and

$$\sum_{4i}' = \sum_{\mu,\nu}(E^2_{\vec{h}_\mu} - 1)(E^2_{\vec{h}_\nu} - 1) \tag{4.46}$$

where $\vec{h}_i = 2\vec{h}_\mu \pm 2\vec{h}_\nu$, and neither $\vec{h}_\mu$ nor $\vec{h}_\nu$ is of the type (h, 0, l) or (0, k, 0), were computed for each of the ten values

Table 14. The signs of ten E_{hkl}'s computed from (4.43).

$\vec{h}$	E^2	Σ'_{1i}	Σ'_{3i}	Σ'_{4i}	Σ'^{*}_{i}	Expected signs (Abrahams et al.)
$102\bar{4}$	8.73	-1.5	- 0.3	-141	-0.25	-
062	7.67	+1.6	-41.0	+ 45	-0.57	-
$66\bar{6}$	7.13	-1.9	-18.8	- 65	-0.48	-
262	5.35	-1.0	-10.1	- 47	-0.27	-
$44\bar{2}$	4.24	-1.8	-16.6	- 17	-0.37	-
026	4.23	-1.4	+37.9	+ 73	+0.69	+
$82\bar{4}$	3.05	-1.8	-13.6	-131	-0.47	-
224	3.00	-1.9	- 3.4	-150	-0.33	-
$62\bar{2}$	3.00	+2.9	+ 3.1	+247	+0.50	+
$24\bar{2}$	2.86	-2.0	-15.6	-136	-0.52	-

$^*\Sigma'_i = 0.0391\,\Sigma'_{1i} + 0.0169\,\Sigma'_{3i} + 0.00132\,\Sigma'_{4i}$.

of i. Here j ranges over the twenty values for which the signs of $E_{h_j 0 l_j}$ (including the incorrect one) have been previously determined. In accordance with Case 2 of Table 10, Case 7 of Table 11, and Case 15 of Table 12, it is to be expected that the sign of $E_{h_i k_i l_i}$ would be the sign of Σ_i'.

The results are shown in Table 14 together with the signs obtained by Abrahams et al. In this case all ten signs were obtained correctly.

Next, the sign of $E_{10.1.\bar{8}}$ which by Theorem 3 may be arbitrarily specified, is taken to be minus in order to compare with that of Abrahams et al. By Theorem 3 the signs of all normalized structure factors E_{hkl}, where (h k l) is linearly dependent modulo 2 on $(10\,1\,\bar{8})$, are then determined. The values of such E_{hkl}'s are arranged in decreasing order and the first ten listed in Table 15. Using (4.07) and (4.08), for each of the nine values of i, i = 2, 3, · · ·, 10, the following functions were computed:

$$\Sigma_i^{ii} = \frac{\sum_{j=1}^{36} Z_j^4}{\left(\sum_{j=1}^{36} Z_j^2\right)^2}\left(\frac{1}{4}\Sigma_{3i}^{iii} + \frac{1}{2}\Sigma_{3i}^{iv}\right), \qquad (4.47)$$

$$\Sigma_{3i}^{iii} = \sum_{\substack{j,\mu \\ j<i}} (-1)^{h_\mu + k_\mu + \epsilon} E_{h_j k_j l_j} (E^2_{h_\mu k_\mu l_\mu} - 1), \qquad (4.48)$$

where

$$h_\mu = h_i \pm h_j, \qquad (4.49)$$

$$l_\mu = l_i \pm l_j, \qquad (4.50)$$

$$k_\mu = k_i \pm k_j, \qquad (4.51)$$

the upper (lower) signs in (4.49) and (4.50) go together, and ϵ is the total number of minus signs appearing in (4.49) and (4.51) for the particular μ value used; and

Table 15. The signs of ten E_{hkl}'s computed from (4.47).

$\vec{h}$	E^2	Σ_i''	Expected signs (Abrahams et al.)
10 1 $\bar{8}$	6.22	-	-
8 3 $\bar{2}$	6.06	-0.13	-
10 1 $\bar{4}$	5.97	-0.71	-
8 1 $\bar{8}$	5.15	-1.36	-
2 7 2	4.61	+0.54	+
2 7 $\bar{2}$	4.42	+0.13	+
8 3 $\bar{4}$	4.35	-1.25	-
4 1 $\bar{2}$	4.00	+0.23	-
2 1 0	3.52	-1.51	-
4 1 $\bar{6}$	3.21	+0.50	+

$$\sum_{3i}^{iv} = \sum_{\substack{j,\mu \\ j<i}} \mp (-1)^{h_\mu + k_\mu} E_{h_j k_j l_j} (E^2_{h_\mu k_\mu l_\mu} - 1) \qquad (4.52)$$

where

$$h_\mu = h_i \pm h_j, \quad k_j = k_i, \quad l_\mu = l_i \pm l_j, \qquad (4.53)$$

so that Σ_{3i}^{iv} exists only if, for at least one value of $j < i$, $k_j = k_i$, and the upper (lower) signs in (4.52) and (4.53) go together. For each fixed value of i, j ranges over the i - 1 preceding values in Table 15 for which the signs have been determined. In accordance with Cases 5 and 8 of Table 11, it is to be expected that the sign of $E_{h_i k_i l_i}$ would be the sign of Σ_i''. The results are shown in Table 15 together with the signs obtained by Abrahams et al. In this case one sign was obtained incorrectly.

It should be noted in Tables 13 and 14 that even the partial sums Σ_{1i}, Σ_{3i}, Σ'_{1i}, Σ'_{3i}, and Σ'_{5i} very often yield the correct signs. The total sums Σ_i, Σ'_i and Σ''_i are themselves only small samples of the complete sums described in the procedure for sign determination previously outlined. It is thus seen that these examples do not constitute a systematic attempt to determine the signs, but rather are an illustration of the application of specific formulas. Nevertheless the results are sufficiently good to indicate that continued application of these formulas to the remainder of the data would result in the correct assignment of the phases of a large number of structure factors. There can be little doubt of the success of the complete procedure.

It should be emphasized that the only information which has been used is the chemical composition of the crystal, the observed x-ray intensities, and the crystal symmetry. No previous knowledge of either phase or structure has been used.

The computations involved in the procedure for sign determination are quite simple and well adapted for IBM techniques. In fact some of them are so simple that they can readily be carried out by means of a hand computer.

Geometric Interpretation

Although the mathematical theory is somewhat involved, the final formulas for phase determination are of an extremely simple nature. One is thus led to expect that a relatively simple interpretation of these formulas exists; and for the space group P$\bar{1}$ this is, in fact, the case. We shall restrict attention to (4.06), although a similar discussion applies to the others. Eq. (4.06) states that if both $E^2_{\vec{h}_\mu}$ and $E^2_{\vec{h}_\nu}$ are large compared to their average values (namely unity), i.e., if both $F^2_{\vec{h}_\mu}$ and $F^2_{\vec{h}_\nu}$ are larger than their average values; or if both $E^2_{\vec{h}_\mu}$ and $E^2_{\vec{h}_\nu}$ are small, then the sign of $E_{\vec{h}}$, where $\vec{h} = 2\vec{h}_\mu + 2\vec{h}_\nu$ is more likely to be plus than minus. On the other hand, if one of $E^2_{\vec{h}_\mu}$, $E^2_{\vec{h}_\nu}$ is large and the other is small, then $E_{\vec{h}}$ is more likely to be minus than plus.

Let us consider the case that $E^2_{\vec{h}_\mu}$ is extremely large. Then the angles $\theta_{\mu j}$ of the $\vec{h}_\mu$ - vector polygon all tend to cluster about 0 or about π as shown in Fig. 2. Likewise, if $E^2_{\vec{h}_\nu}$ is also extremely large then the angles $\theta_{\nu j}$ of the

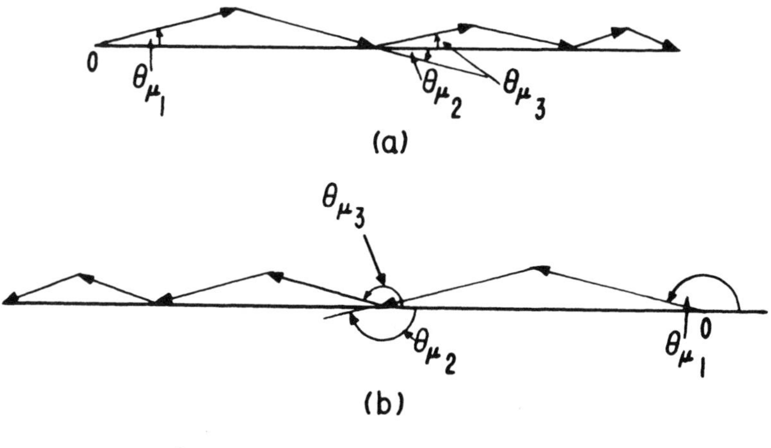

Fig. 2. The $\vec{h}_\mu$ -vector polygon when $E^2_{\vec{h}_\mu}$ is extremely large. $\theta_{\mu j} = -2\pi(hx_j + ky_j + lz_j)$, $\theta_{\mu_1} = -\theta_{\mu_2}$, $\theta_{\mu_3} = -\theta_{\mu_4}$, etc. In (a) the θ_μ cluster about 0, in (b) about π.

$\vec{h}_\nu$ - vector polygon all tend to cluster about 0 or about π. Hence, the values of the angles θ_j of the $\vec{h}$ - vector polygon tend to cluster about one of $2(0 + 0)$, $2(0 + \pi)$, $2(\pi + 0)$, or $2(\pi + \pi)$, i.e., 0 in any case. In short the likelihood is that $E_{\vec{h}}$ will be positive rather than negative, in qualitative agreement with (4.06).

If $E^2_{\vec{h}_\mu}$ is extremely small then the angles θ_{μ_j} of the $\vec{h}_\mu$ - vector polygon tend to cluster about $\pm\frac{\pi}{2}$ (Fig. 3). It should be noted that while values of θ_{μ_j} in the neighborhood of 0 accompanied by θ_{μ_j} values in the neighborhood of π would also give rise to a small value of $E^2_{\vec{h}_\mu}$, this situation is not as probable as the one in which most angles θ_{μ_j} cluster about $\pm\frac{\pi}{2}$. The reason for this is that values of θ_{μ_j} in the neighborhood of 0 require that other values of θ_{μ_j} be

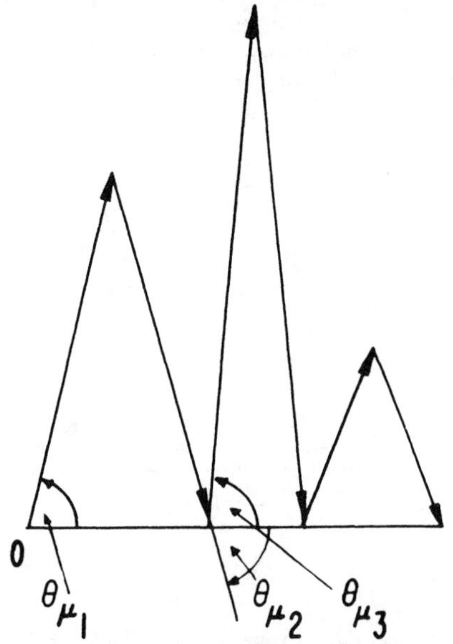

Fig. 3. The $\vec{h}_\mu$ - vector polygon when $E^2_{\vec{h}_\mu}$ is extremely small. $\theta_{\mu_j} = -2\pi(hx_j + ky_j + lz_j)$, $\theta_{\mu_1} = -\theta_{\mu_2}$, $\theta_{\mu_3} = -\theta_{\mu_4}$, etc. The θ_μ cluster about $\pm\frac{\pi}{2}$.

in the neighborhood of π. However values of θ_{μ_j} in the neighborhood of $\frac{\pi}{2}$ are automatically accompanied by values of θ_{μ_j} in the neighborhood of $-\frac{\pi}{2}$ due to the crystal symmetry. Hence it is the latter situation rather than the former which is more likely to give rise to the known small value of $E^2_{\vec{h}\mu}$. The tendency therefore is for the θ_{μ_j} values to cluster about $\pm\frac{\pi}{2}$. Likewise, if $E^2_{\vec{h}\nu}$ is very small, the values of θ_{ν_j} tend to cluster about $\pm\frac{\pi}{2}$. Then the values of θ_j cluster about one of $2\left(\frac{\pi}{2}+\frac{\pi}{2}\right)$, $2\left(-\frac{\pi}{2}+\frac{\pi}{2}\right)$, $2\left(\frac{\pi}{2}-\frac{\pi}{2}\right)$, $2\left(-\frac{\pi}{2}-\frac{\pi}{2}\right)$, i.e., 0 in any case. Again the sign of $E_{\vec{h}}$ is seen to be more likely plus than minus, in agreement with (4.06).

Finally, if one of $E^2_{\vec{h}\mu}$, $E^2_{\vec{h}\nu}$ is extremely large and the other extremely small then the values of $\theta_{\vec{h}\mu}$ tend to cluster about 0 or π and the values of $\theta_{\vec{h}\nu}$ tend to cluster about $\pm\frac{\pi}{2}$, or vice versa. In any case the values of $\theta_{\vec{h}}$ tend to cluster about one of $2\left(0+\frac{\pi}{2}\right)$, $2\left(0-\frac{\pi}{2}\right)$, $2\left(\pi+\frac{\pi}{2}\right)$, $2\left(\pi-\frac{\pi}{2}\right)$, i.e. π. Hence the sign of $E_{\vec{h}}$ is more likely to be minus than plus in this case, again in agreement with (4.06).

Chapter 5

SPECIAL POSITIONS

Probabilities

So far, this paper has been concerned with crystals having atoms only in general positions. However, the methods developed apply equally well to crystals having atoms in special positions in addition to those in general positions. Equation (1.01) is replaced by the more general

$$F = \sum_{j=1}^{t} f_j \xi_j, \quad t = \sum_{i=1}^{v} N_i/n_i, \tag{5.01}$$

where v is the total number of types of positions (special and general), N_i is the number of atoms in each type of position and n_i is the number of equivalent atoms in the corresponding type. The functions ξ_j maintain the same form for each fixed value of i, i.e., for values of j corresponding to a fixed type of position, and, together with v and n_i, are known for each space group. The joint probability that $\xi_{j\mu}$ (Eq. (3.01)) lie in the interval $\xi_{j\mu}$, $\xi_{j\mu} + d\xi_{j\mu}$, for $\mu = 1, 2, \cdots, m$, now depends on j and is denoted by $p_j(\xi_{j1}, \cdots, \xi_{jm})$. Let $P_1(A_1, \cdots, A_m)\, dA_1 \cdots dA_m$ be the joint probability that $F_{\vec{h}_\mu}$ lie in the interval A_μ, $A_\mu + dA_\mu$, $\mu = 1, 2, \cdots, m$. Just as in the proof of (3.02) and (3.03), we have the fundamental result

$$P_1(A_1, \cdots, A_m) = \frac{1}{(2\pi)^m} \int_{-\infty}^{\infty} \cdots \int_{-\infty}^{\infty} \exp\left(-i \sum_{\mu=1}^{m} A_\mu w_\mu\right) \times$$

$$\prod_{j=1}^{t} q_j(f_{j1} w_1, \cdots, f_{jm} w_m) dw_1 \cdots dw_m, \tag{5.02}$$

where

$$q_j(f_{j1}w_1, \cdots, f_{jm}w_m) = \int_{-\infty}^{\infty} \cdots \int_{-\infty}^{\infty} p_j(\xi_{j1}, \cdots, \xi_{jm}) \times$$

$$\exp\left(i \sum_{\mu=1}^{m} f_{j\mu}\, \xi_{j\mu}\, w_\mu\right) d\xi_{j1} \cdots d\xi_{jm}, \qquad (5.03)$$

and $f_{j\mu}$ is given by (3.04). As before it is not necessary to know the functions $p_j(\xi_{j1}, \cdots, \xi_{jm})$. Only the mixed moments

$$m_{j\lambda_1\lambda_2 \cdots \lambda_m} = \int_{-\infty}^{\infty} \cdots \int_{-\infty}^{\infty} p_j(\xi_{j1}, \cdots, \xi_{jm}) \times$$

$$\xi_{j_1}^{\lambda_1}\, \xi_{j_2}^{\lambda_2} \cdots \xi_{jm}^{\lambda_m}\, d\xi_{j1}\, d\xi_{j2} \cdots d\xi_{jm} \qquad (5.04)$$

$$= \int_0^1\int_0^1\int_0^1 \xi_{j1}^{\lambda_1}(x,y,z,\vec{h}_1)\, \xi_{j2}^{\lambda_2}(x,y,z,\vec{h}_2) \cdots$$

$$\cdots \xi_{jm}^{\lambda_m}(x,y,z,\vec{h}_m)\, dx\,dy\,dz \qquad (5.05)$$

are needed and the equivalence of these two expressions for $m_{j\lambda_1\lambda_2 \cdots \lambda_m}$ may be inferred by interpreting the mixed moment as an average or expected value of $\xi_{j1}^{\lambda_1}\, \xi_{j2}^{\lambda_2} \cdots \xi_{jm}^{\lambda_m}$.

As in the proof of (3.14), we find for the case m = 2

$$P_1(A_1, A_2) =$$

$$\frac{\exp\left(-\dfrac{A_1^2}{2\sum_{i=1}^{v}\left(\dfrac{m_{i20}}{n_i}\sum_{j=1}^{N_i} f_{j1}^2\right)} - \dfrac{A_2^2}{2\sum_{i=1}^{v}\left(\dfrac{m_{i02}}{n_i}\sum_{j=1}^{N_i} f_{j2}^2\right)}\right)}{2\pi\left(\sum_{i=1}^{v}\left(\dfrac{m_{i20}}{n_i}\sum_{j=1}^{N_i} f_{j1}^2\right)\sum_{i=1}^{v}\left(\dfrac{m_{i02}}{n_i}\sum_{j=1}^{N_i} f_{j2}^2\right)\right)^{\frac{1}{2}}} \times$$

$$\left\{1+\frac{\sum_{i=1}^{v}\left(\frac{m_{i12}}{n_i}\sum_{j=1}^{N_i} f_{j1}\, f_{j2}^2\right)}{2\left(\sum_{i=1}^{v}\left(\frac{m_{i20}}{n_i}\sum_{j=1}^{N_i} f_{j1}^2\right)\right)^{\frac{1}{2}}\left(\sum_{i=1}^{v}\left(\frac{m_{i02}}{n_i}\sum_{j=1}^{N_i} f_{j2}^2\right)\right)} \times \right.$$

$$\left.\frac{A_1}{\left(\sum_{i=1}^{v}\left(\frac{m_{i20}}{n_i}\sum_{j=1}^{N_i} f_{j1}^2\right)\right)^{\frac{1}{2}}}\left(\frac{A_2^2}{\sum_{i=1}^{v}\left(\frac{m_{i02}}{n_i}\sum_{j=1}^{N_i} f_{j2}^2\right)} - 1\right)\right\}, \quad (5.06)$$

where i ranges through the v different types of position and, for each i, j ranges through all N_i atoms in the corresponding type. Now we define the <u>normalized structure factor</u> $E_{\vec{h}}$ by means of

$$E_{\vec{h}} = \frac{F_{\vec{h}}}{\left(\sum_{i=1}^{v}\left(\frac{m_{i2}}{n_i}\sum_{j=1}^{N_i} f_j^2\right)\right)^{\frac{1}{2}}}, \quad (5.07)$$

an obvious generalization of (3.15). As before, $E_{\vec{h}}$ and $F_{\vec{h}}$ have the same sign.

Assuming that the atoms range uniformly and independently throughout the unit cell (subject to the restraints imposed by symmetry)

$$\langle F_{\vec{h}}^2\rangle_{\vec{r}} = f'^{\,2} + \sum_{i=1}^{v_n}\left(\frac{m_{i2}}{n_i}\sum_{j=1}^{N_i} f_j^2\right) \quad (5.08)$$

(Karle and Hauptman, (1953 a), footnote, p. 135) where f' is the contribution to the structure factor of those atoms in the fixed special positions, i ranges over the $v_n = v - v_f$ types of special positions, exclusive of the fixed special positions, and v_f is the number of types of fixed special positions. For each i, m_{i2} is a function of $\vec{h}$ but takes on only a finite number of different values. The indices $\vec{h}_1$ and $\vec{h}_2$ will be called similar relative to the i type of special position if m_{i2} has the same value for $\vec{h}_1$ and $\vec{h}_2$.

Furthermore, it is readily verified that

$$<f'^2>_{\vec{h}} = \sum_{i=1}^{v_f} \left(\frac{m_{i_2}}{n_i} \sum_{j=1}^{N_i} f_j^2 \right) \qquad (5.09)$$

where $<f'^2>_{\vec{h}}$ is the average value of f'^2 as $\vec{h}$ ranges uniformly over those indices any two of which are similar relative to every type i of fixed special position; i.e., for which, for each fixed value of i, m_{i_2} has a constant value (independent of $\vec{h}$, but depending on i). For each of these sets we obtain, in general, a different value of $<f'^2>_{\vec{h}}$. It is known that if the crystal contains no atom in a fixed special position, averaging over $\vec{h}$ is equivalent, in general, to averaging over $\vec{r}$. Hence if we use (5.09) to re-write (5.08), and interpret $<F^2>_{\vec{h}_i}$ as the average over any set of $\vec{h}_i$ giving rise to a constant value of m_{i_2} for each fixed i (but different for different i's), we get

$$<F^2>_{\vec{h}} = \sum_{i=1}^{v} \left(\frac{m_{i_2}}{n_i} \sum_{j=1}^{N_i} f_j^2 \right). \qquad (5.10)$$

Equations (5.07) to (5.10) lead to the generalization of (3.18),

$$<E^2>_{\vec{h}} = 1 \qquad (5.11)$$

where $\vec{h}$ ranges uniformly over those indices for which m_{i_2} has a constant value for each fixed i, or even over all the indices. The importance of (5.11) is that, like (3.18), it leads to a procedure for treating the experimental data previous to their application in the phase determining relationships. It should be noted however that, owing to (5.08), $<E_{\vec{h}}^2>_{\vec{r}}$ is not necessarily equal to unity if the crystal contains atoms in fixed special positions.

Analysis of Data

If no extinctions occur, other than those associated with one and two dimensional reflections, it is possible to adjust the intensity data for vibrational motion and absolute scale by Wilson's method (1949), e.g. (Karle and Hauptman, 1953b), and use the phase determining procedure outlined for atoms in general positions even though atoms occur in special positions also. It is only when different extinctions occur for the various types of positions that the determination of $E_{\vec{h}}$, as defined by (5.07), does not reduce to the previous case. In general, for each space group, the various types of positions may be grouped into classes characterized by the conditions limiting reflections and by the values of m_{i2}. Such a situation is typified by the space group Pnnn when all types of positions are occupied. A possible procedure for handling the experimental data for this space group will now be outlined and a similar treatment applies generally.

We restrict our attention at first to those intensities for which $h\,k\,l \neq 0$. There are thirteen types which fall into three classes:

Class 1: Type m, $\frac{m_{i2}}{n_i} = 1$, $i = 1$;

Class 2: Types a - d, g - l, $\frac{m_{i2}}{n_i} = 2$, $i = 2, 3, \cdots, 11$;

Class 3: Types e, f, $\frac{m_{i2}}{n_i} = 4$, $i = 12, 13$.

We consider the four sets:

Set 1: $h + k + l$ odd, $h + k$ or $k + l$ or $l + h$ odd;

Set 2: $h + k + l$ odd, $h + k$ and $k + l$ and $l + h$ even;

Set 3: $h + k + l$ even, $h + k$ or $k + l$ or $l + h$ odd;

Set 4: $h + k + l$ even, $h + k$ and $k + l$ and $l + h$ even.

For small but statistically significant $s = \frac{\sin\theta}{\lambda}$ intervals, the average of the squares of the structure factors (corrected experimental intensities) I_1, I_2, I_3, I_4 is determined for each of these four sets of indices. From (5.10)

$$\left.\begin{aligned}
&\sum_{j=1}^{N_1} f_j^2 = K\,I_1 \\
&\sum_{j=1}^{N_1} f_j^2 + 4\sum_{i=12}^{13}\sum_{j=1}^{N_i} f_j^2 = K\,I_2 \\
&\sum_{j=1}^{N_1} f_j^2 + 2\sum_{i=2}^{11}\sum_{j=1}^{N_i} f_j^2 = K\,I_3 \\
&\sum_{j=1}^{N_1} f_j^2 + 2\sum_{i=2}^{11}\sum_{j=1}^{N_i} f_j^2 + 4\sum_{i=12}^{13}\sum_{j=1}^{N_i} f_j^2 = K\,I_4
\end{aligned}\right\} \quad (5.12)$$

where K is a constant. Since $\sum_{i=1}^{13}\sum_{j=1}^{N_i} f_j^2$ is known from the chemical composition, the values of $\sum_{j=1}^{N_1} f_j^2$, $\sum_{i=2}^{11}\sum_{j=1}^{N_i} f_j^2$, and $\sum_{i=12}^{13}\sum_{j=1}^{N_i} f_j^2$ are easily found from (5.12). The analysis of those intensities for which h k l = 0 may be carried out in a similar manner. In this way we find the information concerning the chemical composition of the various types of position which is required by (5.07). Using (5.10), in a manner analogous to the procedure outlined by Karle and Hauptman (1953b) the experimental intensities may be corrected for vibrational motion and placed on an absolute scale. Hence the values of $E_{\vec{h}}$ needed in the phase determining formulas outlined below may be obtained from (5.07).

In terms of the normalized structure factors (5.07), Eq. (5.06) becomes the generalization of (3.19),

$$P(E_1,E_2) = \frac{\exp\left(-\frac{1}{2}E_1^2 - \frac{1}{2}E_2^2\right)}{2\pi}\left\{1 + \frac{\sum_{i=1}^{v}\left(\frac{m_{i12}}{n_i}\sum_{j=1}^{N_i} f_{j1} f_{j2}^2\right) E_1(E_2^2 - 1)}{2\left(\sum_{i=1}^{v}\left(\frac{m_{i20}}{n_i}\sum_{j=1}^{N_i} f_{j1}^2\right)\right)^{\frac{1}{2}}\left(\sum_{i=1}^{v}\left(\frac{m_{i02}}{n_i}\sum_{j=1}^{N_i} f_{j2}^2\right)\right)}\right\} \tag{5.13}$$

where $P(E_1, E_2)\, dE_1\, dE_2$ is the probability that both $E_{\vec{h}_1}$ lies between E_1 and $E_1 + dE_1$ and $E_{\vec{h}_2}$ lies between E_2 and $E_2 + dE_2$. The analogous generalizations of (3.20) to (3.25) are now self-evident and no further elaboration is needed. It has already been seen how (3.19) to (3.25) lead to a procedure for phase determination. In the same way the generalizations of these equations lead to a procedure for phase determination valid for crystals having atoms in special positions as well as in general positions. The details of this analysis are similar to those for general positions and are therefore omitted.

Chapter 6

CONCLUDING REMARKS

Homometric Structures

Throughout this Monograph it has been tacitly assumed that the phase problem has a unique solution. However, the solution to the phase problem is not necessarily unique (Patterson, 1944). For simplicity let us assume that there are just two centrosymmetric homometric structures S_1 and S_2, i.e. two solutions to the phase problem. It is then clear that any phase which is a structure seminvariant either has the same value for S_1 and S_2 or else has one value for S_1 and another for S_2. The totality of phases which are structure seminvariants is then divided into two classes, those having the same value for the two possible structures and those having different values for the two structures. It is to be expected that each seminvariant belonging to the first class, C_s, will be correctly determined by the procedure since it has the same value for both structures. However the procedure should yield indecisive results for any seminvariant belonging to the second class, C_d, for the value of such a phase is different for the different structures. It appears then that if we specify arbitrarily the value of one phase $\phi_{\vec{h}}$ belonging to C_d, thus distinguishing between the two structures S_1 and S_2, all the remaining phases belonging to C_d will be determined; and it is to be expected that the procedure will determine these correctly. Since the value of $\phi_{\vec{h}}$ may be specified in two different ways we obtain in this way the two solutions of the phase problem.

Although space does not permit a detailed study of the problems which arise when there are more than two solutions of the phase problem it appears likely, in view of the previous discussion, that the probability methods developed here will be capable of revealing the presence of multiple solutions, and then of finding all solutions, i.e. all homometric structures, in any given case. The very brief discussion of the previous paragraph already indicates in fact how the homometric structures are likely to reveal themselves, and one is almost tempted to conjecture that the number of centrosymmetric homometric structures is always a power of two, say 2^k. The values of k phases which are seminvariants may then be arbitrarily specified. Whether the probability methods are sufficiently powerful to cope with this problem and others of a general nature arising from the presence of homometric structures, remains an unanswered question.

Summary

Probability distributions have been obtained for a structure factor based upon the knowledge of certain sets of intensities or phases. The probability that a structure factor be positive may then be inferred from these distributions. The solution of the phase problem based on these results leads to a routine and practical scheme for phase determination which is valid for all centrosymmetric space groups and which requires no a priori knowledge of any phase. Only routine, but often tedious, mathematical computations are required to determine once and for all a specific procedure to be used for each centrosymmetric space group. These have been carried out in detail for space group $P\bar{1}$ and, to a lesser extent, for $P2_1/a$. The procedure is readily adapted for IBM techniques and may, in fact, often be carried out in part by means of a hand computer. The solution requires a knowledge only of a sufficient number of x-ray intensities and of the chemical composition of the crystal. These methods appear to be useful for even the most complex crystals, provided only that the number of observed x-ray intensities is sufficiently large. The analysis includes the case that the crystal contains atoms in special positions as well as in general positions. Illustrative examples have been worked out in detail.

APPENDIX

DISTRIBUTION OF THE FRACTIONAL PART OF hx + ky + lz

For a fixed crystal structure the frequency distribution of the structure factors F_{hkl} is determined by permitting the indices h,k,l to range uniformly and independently over the integers. It has been shown by Weyl (1915-16) that if the numbers x,y,z are rationally independent (i.e. no integers m_1, m_2, m_3, m, not all zero, exist satisfying $m_1x + m_2y + m_3z = m$) then the fractional part of hx + ky + lz is uniformly distributed as h,k,l range uniformly and independently over the integers. Since the structure factor F_{hkl} is a function of the fractional part of hx + ky + lz, this result of Weyl affords the possibility of determining theoretically the frequency distribution of the structure factors for any fixed structure, as pointed out in the Preface. In fact we shall show that if x,y,z are chance variables, and x is uniformly distributed in the interval (0,1), but occurs with zero probability elsewhere, then the fractional part of hx + ky + lz is also uniformly distributed in the interval (0,1) provided that h is an integer. This result is one reason for the underlying assumption (in the second Section of Chapter 1) that, in the absence of any observed intensities, the atoms in the asymmetric unit of a crystal are uniformly and independently distributed. It should be noted that if no one of h,k,l is an integer, the fractional part of hx + ky + lz will, in general, not be uniformly distributed in the interval (0,1) even if x,y, and z are uniformly and independently distributed in the interval (0,1).

Assume x to be uniformly distributed in the interval (0,1) and ky to have a probability distribution f(y). Then the probability distribution function of hx is 1/h in the interval (0,h) and is zero elsewhere, while that of u = hx + ky is given by (Uspensky, p. 269)

$$g(u) = \frac{1}{h} \int_0^h f(u - y)\, dy. \tag{7.1}$$

Denote the fractional part of u by η. Then η will lie in the interval $(\eta, \eta + d\eta)$ if and only if u lies in one of the intervals

$$(j + \eta,\ j + \eta + d\eta), \qquad j = 0, \pm 1, \pm 2, \cdots. \tag{7.2}$$

Since the cases (7.2) are mutually exclusive and exhaustive, the theorem of total probability applies (Uspensky, p. 27) and the probability distribution of the fractional part η of hx + ky is

$$H(\eta) = \sum_{j=-\infty}^{\infty} g(j + \eta) \tag{7.3}$$

$$= \sum_{j=-\infty}^{\infty} \frac{1}{h} \int_0^h f(j + \eta - y)\, dy \tag{7.4}$$

$$= \frac{1}{h} \sum_{j=-\infty}^{\infty} \int_{\eta - h}^{\eta} f(j + w)\, dw, \tag{7.5}$$

and, on the assumption that the summation and integration may be interchanged,

$$H(\eta) = \frac{1}{h} \int_{\eta - h}^{\eta} \sum_{j=-\infty}^{\infty} f(j + w)\, dw. \tag{7.6}$$

In view of

$$\sum_{j=-\infty}^{\infty} f(j + w + 1) = \sum_{j=-\infty}^{\infty} f(j + w), \tag{7.7}$$

it follows that $\sum_{j=-\infty}^{\infty} f(j + w)$ is a periodic function of w of period 1. The derivative of (7.6) is, in view of (7.7),

$$\frac{d}{d\eta} H(\eta) = \frac{1}{h}\left[\sum_{j=-\infty}^{\infty} f(j+\eta) - \sum_{j=-\infty}^{\infty} f(j+\eta-h)\right] = 0, \qquad (7.8)$$

since h is an integer. It is clear therefore that $H(\eta)$ is independent of η so that η is uniformly distributed in the interval (0,1). Since the fractional part of hx + ky + lz is equal to the fractional part of η + lz, the same argument shows that the fractional part of hx + ky + lz is uniformly distributed in the interval (0,1).

REFERENCES

Abrahams, S. C., Robertson, J. M., and White, J. G. (1949). Acta Cryst. 2, 233.

Cochran, W. (1952). Acta Cryst. 5, 65.

Harker, D. and Kasper, J. (1948). Acta Cryst. 1, 70.

Hauptman, H. and Karle, J. (1952). Acta Cryst. 5, 48.

Hauptman, H. and Karle, J. (1953). Acta Cryst. 6, 136.

International Tables for X-Ray Crystallography, Vol. I, (1952), The Kynoch Press, Birmingham, England.

Karle, J. and Hauptman, H. (1950). Acta Cryst. 3, 181.

Karle, J. and Hauptman, H. (1953a). Acta Cryst. 6, 131.

Karle, J. and Hauptman, H. (1953b). Acta Cryst. 6, 473.

Okaya, Y. and Nitta, I. (1952). Acta Cryst. 5, 564.

Patterson, A. L. (1944). Phys. Rev. 64, 195.

Sayre, D. (1952). Acta Cryst. 5, 60.

Uspensky, J. V. (1937). Introduction to Mathematical Probability, McGraw Hill, New York.

Weyl, H. (1915-16). Math. Ann. 77, 313.

Wilson, A. J. C. (1949). Acta Cryst. 2, 318.

Zachariasen, W. H. (1952). Acta Cryst. 5, 68.

INDEX